GC/MS

GC/MS
A Practical User's Guide

MARVIN McMASTER
AND
CHRISTOPHER McMASTER

 **WILEY-VCH**

New York / Chichester / Weinheim / Brisbane / Singapore / Toronto

Library of Congress Cataloging-in-Publication Data:

McMaster, Marvin C.
    GC/MS : a practical user's guide / by Marvin McMaster and
Christopher McMaster.
        p.   cm.
    Includes index.
    ISBN 0-471-24826-6 (cloth : alk. paper)
    1. Gas chromatography.   2. Mass spectrometry.   I. McMaster,
Christopher.   II. Title.
    QD79.C45M423   1998
    543'.0873—dc21   97-48529

CONTENTS

PART II. GC/MS OPTIMIZATION   59

PART III. SPECIFIC APPLICATIONS OF MASS SPECTROMETRY   97

PREFACE

This handbook is presented in sections because I believe it is easier to learn that way. Part I first presents a comparative look at gas chromatography/mass spectrometry (GC/MS) and competitive instrumentation. Then an overview of the components of a generic GC/MS system is provided. Finally, we walk you through getting a system set up and performing an analysis that provides the information you are seeking.

After some hands-on experience, Part II on optimization provides information on tuning and calibration, cleaning, troubleshooting, processing data, and interfacing to other analytical and data systems, that is, getting the system up and running, keeping it up and running, and getting the information where it is needed.

Part III provides information on the use of GC/MS systems in research and environmental laboratories. Although quadrupole mass spectrometers predominate in commercial laboratories, there is growing use for ion traps, time of flight, and triple-quadrupole MS/MS systems in research laboratories, and these are discussed briefly. Magnetic sector systems, which dominated the early growth of mass spectrometry, are making a resurgence in the accurate mass determination required for the molecular formula and structure reporting in chemical publications, and these are discussed next.

I hope you enjoy the book and find it of use in your laboratory. When I was designing my course at the University of Missouri at St. Louis, I looked for a textbook to use. I could not find a practical guide to using and maintaining a GC/MS system—nothing that surveyed modern systems and applications.

MARVIN C. MCMASTER

GC/MS

PART I

A GC/MS PRIMER

CHAPTER 1

INTRODUCTION

The combination of gas chromatography (GC) for separation and mass spectrometry (MS) for detection and identification of the components of a mixture of compounds is rapidly becoming the definitive analytical tool in the research and commercial analytical laboratory. The GC/MS systems come in many varieties and sizes depending on the work they are designed to accomplish. Since the most common analyzer used in modern mass spectrometers is the quadrupole, we will focus on this means of separating ion fragments of different masses. Discussion of ion trap, time of flight, Fourier transform mass spectrometry (FTMS), and magnetic sector instruments will be reserved for later sections in the book.

The quadrupole operational model is the same for benchtop production units and for floor-standing research instruments. The actual analyzer has changed little in the last 10–12 years except to grow smaller in size. High-vacuum pumping has paralleled the changes in the analyzer, especially in the high-efficiency turbo pumps that have been reduced to the size of a large fist. Sampling and injection techniques have improved gradually over the last few years.

The most dramatic changes have been in the area of control and processing software and data storage capability. In the last 5 years accelerating computer technology has reduced the computer hardware and software systems shipped with the original system to historical oddities. In the face of newer, more powerful, easier to use computer systems the older DEC 10, RTE (a Hewlett Packard mini computer GS/MS control system), and Pascal-based control and data processing systems seem to many operators to be lumbering, antiquated monstrosities.

The two most common reasons given for replacing a GC/MS system are the slow processing time and the cost of operator training. This is followed by unavailability of replacement parts as systems are discontinued by manufacturers. The inability of software to interface with and control modern gas chromatographic and sample preparation systems is the final reason given for replacement.

Seldom, if ever, is the complaint that the older systems do not work, or that they give incorrect values. In many cases, according to users who run both types of instruments, the older hardware systems appear better built and more stable in day-to-day operation than new models. Many require less cleaning and maintenance. This has led to a growing market for replacement data acquisition and processing systems. Where possible the control system should also be updated, allowing access to modern auxiliary equipment and eliminating the necessity for coordinating dual computers of differing ages and temperaments.

Replacement of older systems with the newest processing system on the market is not without its problems. Fear of loss of access to archived data stored in outdated, proprietary data formats is a common worry of laboratories doing commercial analysis.

1.1. WHY USE GC/MS?

Gas chromatography is a popular, powerful, reasonably inexpensive, and easy-to-use analytical tool. Mixtures to be analyzed are injected into an inert gas stream and swept into a tube packed with a solid support coated with a resolving liquid phase. Absorptive interaction between the components in the gas stream and the coating leads to a differential separation of the components of the mixture, which are then swept in order through a detector flow cell. Gas chromatography suffers from a few weaknesses, such as its requirement for volatile compounds, but its major problem is the lack of definitive proof of the nature of the detected compounds as they are separated. For most GC detectors, identification is based solely on retention time on the column. Since many compounds may possess the same retention time, we are left in doubt as to the nature and purity of the compound(s) in the separated peak.

The mass spectrometer takes injected material, ionizes it in a high vacuum, propels and focuses these ions and their fragmentation products through a magnetic mass analyzer, and then collects and measures the amounts of each selected ions in a detector. A mass spectrometer is an excellent tool for clearly identifying the structure of a single compound but is less useful when presented with a mixture.

The combination of the two components into a single GC/MS system forms an instrument capable of separating mixtures into their individual components, identifying and then providing quantitative and qualitative information on the amounts and chemical structure of each compound. It still possesses the weaknesses of both components. It requires volatile components, and because of this requirement, it has some molecular weight limits. The mass spectrometer must be tuned and calibrated before meaningful data can be obtained. The data produced have time, intensity, and spectral components and require a computer with a large storage system for processing and identifying the components. A major drawback of the system is that it is very expensive compared to other analytical systems. With continued improvement of the system, hopefully the cost will be lowered because this system and/or the liquid chromatograph/mass spectrometer belongs on every laboratory benchtop used for organic or biochemical synthesis and analysis.

Determination of the molecular structure of a compound from its molecular weight and fragmentation spectra is a job for a highly trained specialist. It is beyond the scope and intent of this book to train you in the interpretation of compound structures. Anyone interested in pursuing that goal should work through McLafferty's book in Appendix D, then practice, practice, practice. A later chapter is included to provide tools to evaluate compound assignments in spectral databases. It uses many of the tools employed in interpretation, but its intent is to provide a quick check on the validity of an assignment.

1.2. INTERPRETATION OF FRAGMENTATION DATA VERSUS SPECTRAL LIBRARY SEARCHING

How do we go about extracting meaningful information from spectra and identifying the compounds we have separated? A number of libraries of printed and computerized spectral databases are available. We can use these spectra to compare both masses of fragments and their intensities. Once a likely match is found, we can obtain and run the same compound on our instrument to confirm the identity both by GC retention time and mass spectra. This matching is complicated by the fact that the listed library spectra are run on a variety of types of mass spectrometers and under dissimilar tuning conditions. However, with modern computer database searching techniques, large numbers of spectra can be searched and compared in a very short time. This allows an untrained spectroscopist to use GC/MS for compound identification with some confidence. Using these spectra, target mass fragments characteristic of each compound can be selected that allows its identification among similarly eluting compounds in the chromatogram.

Once compounds have been identified, they can be used as standards to carry out quantitative analysis of mixtures of compounds. Unknown compounds found in quantitation mixtures can be flagged and identified by spectral comparison using library searching. Spectra from scans at chromatography peak fronts and tails can be used to confirm purity or identify the presence of impurities.

1.3. THE GAS CHROMATOGRAPH/MASS SPECTROMETER

From the point of view of the chromatographer the gas chromatograph/ mass spectrometer is simply a gas chromatograph with a very large and very expensive detector but one that can give a definitive identification of the separated compounds. The sample injection and the chromatographic separation are handled exactly the same as in any other analyses. You still get a chromatogram of the separated components at the end. It is what can be done with the chromatographic data that distinguishes the mass spectral detector from an electron capture or a flame ionization detector.

The mass spectrometrist approaches GC/MS from a different point of view. The mass spectrum is everything. The gas chromatograph exist only to aid somewhat in improving difficult separations of compounds with similar mass fragmentations. The true art is the interpretation of spectra and identification of molecular structure and molecular weight.

The truth, of course, lies somewhere in between. A good chromatographic separation based on correct selection of injector type and throat material, column support, carrier gas, and oven temperature ramping and a properly designed interface feeding into the ion source can make or break the mass spectrometric analysis. Without a properly operating vacuum system, ion focusing system, mass analyzer, and detector, the best chromatographic separation in the world is just a waste of the operator's time. It is important to understand the components that make up all parts of the GC/MS system in order to keep the system up and running and performing in a reproducible manner.

1.3.1. A Model of the GC/MS System

There are a number of different possible GC/MS configurations, but all share similar components. There must be some way of getting the sample into the chromatograph, an *injector*. This may or may not include sample purification or preparation components. There must be a *gas chromatograph,* with its carrier gas source and control valving, its temperature control oven, and tubing to connect the injector to the column and out to the mass spectrometer interface. There must be a *column* packed with support and coated with a stationary phase in which the separation occurs. There must be an *interface* module in which the separated compounds are transferred to the mass spectrometer's ionization source without remixing. There must be the *mass spectrometer* system, made up of the ionization source, focusing lens, mass analyzer, detector, and multistage pumping. Finally there must be a *data/control* system to provide mass selection, lens and detector control, and data processing (see Figure 1.1).

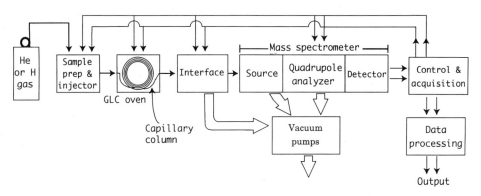

Figure 1.1   A typical GC/MS system diagram.

The injector may be as simple as a septum port on top of the gas chromatograph through which a sample is injected using a graduated capillary syringe. In some cases, this injection port is equipped with a trigger that can start the oven temperature ramping program and/or send a signal to the data/control system to begin acquiring data. For more complex or routine analysis, injection can be made from an autosampler allowing multiple vial injections and automated chromatography and data processing. For crude samples that need preinjection processing there are split/splitless injectors, throat liners with different surface geometry, purge and trap systems, head space analyzers, and cartridge purification systems. All provide sample extraction, cleanup, or volatilization prior to the sample being introduced into the gas chromatographic column.

The gas chromatograph (Figure 1.2) is a temperature-controlled oven designed to hold and heat the GC column. Carrier gas, usually nitrogen, helium, or hydrogen, is used to sweep the injected sample onto and down the column where the separation occurs and then out into the mass spectrometer interface.

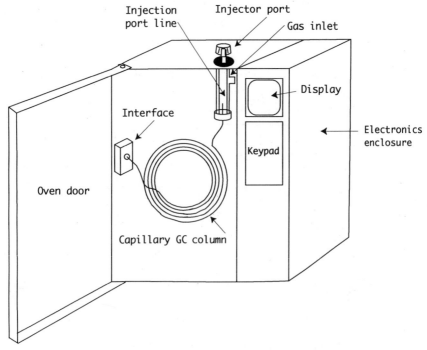

Figure 1.2 Gas chromatograph.

The interface may serve only as a transfer line to carry the pressurized GC output into the evacuated ion source of the mass spectrometer. It can also serve as a sample concentrator by eliminating much of the carrier gas. It can permit carrier gas displacement by a second gas that is more compatible with the desired analysis, that is, carbon dioxide for chemically induced (CI) ionization for molecular weight analysis. It can be used to split the GC output into separate streams that can be sent to different detectors for simultaneous analysis by completely different methods.

The mass spectrometer has three basic sections: an ionization chamber, the analyzer, and the ion detector (Figure 1.3).

In the evacuated ionization chamber the sample is bombarded with electrons or charged molecules to produce ionized sample molecules. These are swept into the high-vacuum analyzer where they are focused electrically and then selected in the quadrupole. The electrically charged poles of the quadrupole create a standing magnetic field in which the ions are aligned. Individual masses are selected from this field by sweeping it with a radio frequency signal. As different frequencies are reached, different mass–charge ratio (m/z) ions are able to escape the analyzer and reach the detector. By sweeping from higher to lower frequency, the available range of m/z ions are released one at a time to the detector, producing a mass spectrum.

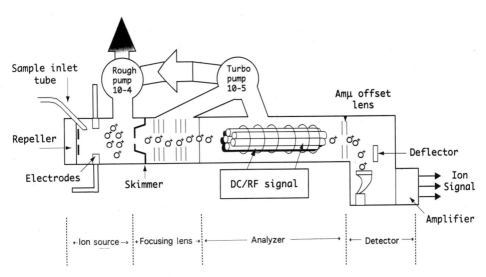

Figure 1.3 Quadrupole mass spectrometer.

On entering the detector, the ions are deflected into a cascade plate where the signal is multiplied and then sent to the data system as an ion current versus m/z versus time. The summed raw signal can be plotted against time as a total-ion chromatogram (TIC) or a single ion m/z can be extracted and plotted against time as a single-ion chromatogram (SIC). At a single time point, the ion current strength for each detected ion fragment can be extracted and plotted over an m/z range producing a mass spectrum. It is important always to remember that the data block produced is three dimensional: m/z versus signal strength versus time. In most other detectors the output is simply signal strength versus time.

1.3.2. A Column Separation Model

Separation of individual compounds in the injected sample occurs in the chromatographic column. The typical gas chromatographic column used for GC/MS is a long, coiled capillary tube of silica with an internal coating of either a viscous liquid such as Carbowax or a wall-bonded organic material.

The injected sample in the carrier gas interacts with this stationary organic phase, and an equilibrium is established between the concentration of each component in the gaseous and solid phases. As fresh carrier gas flushes down the column, each compound comes off the stationary phase at its own rate. Separations increase after many interactions down the length of the column; then each volatile component comes off the column end and into the interface (Figure 1.4).

Since not all compounds are volatile at room temperature, both the injector and the column can be heated to aid in their removal. The column oven allows programmed gradient heating of the column. Temperatures above 400°C are avoided to prevent sample thermal degradation.

Moving down the column, the injection mixture interacts with the packing. Separation is countered by diffusion and wall interactions. Finally, each compound emerges into the interface as a concentration disk, tenuous at first, then rising to a concentration maxima, and then dropping rapidly as the last molecules come off. If we were to run this effluent in an ultraviolet (UV) detector, we would see a rapidly rising peak reach its maximum height and then fall again with a slight tail.

Ideally, each compound emerges as a disk separated from all other disks. In actual separations of real samples, perfect separation is rarely achieved. Compounds of similar chemical structure and physical solubilities are only poorly resolved and coelute. In a chromatographic detector they appear as

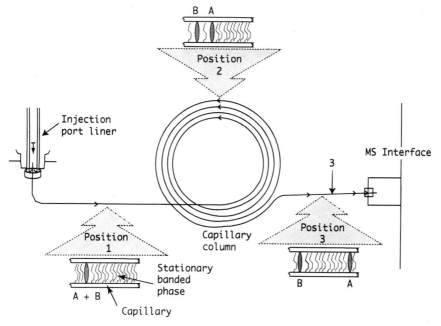

Figure 1.4   Chromatographic column separation model.

overlapping or unresolved peaks. Something else must be done to prove them present, identify their structure, and quantitate the amounts of each compound.

1.3.3. GC/MS Data Models

The simplest data output from the mass spectrometer analyzer is a measurement of total ion current strength versus time, a total-ion chromatogram (TIC), (Figure 1.5).

This is basically a chromatographic output representing a summation of all the ions produced by the mass spectrometer at a given time. The chromatogram produced is similar in appearance to a UV chromatogram, with peaks representing the chromatographic retention of each component present. In a UV detector, however, you would see only the compounds that absorb UV light at the selected wavelength. In the mass spectrometer, any compound capable of being ionized and forming fragments would be detected. The mass spectrometer serves as a universal chromatographic detector.

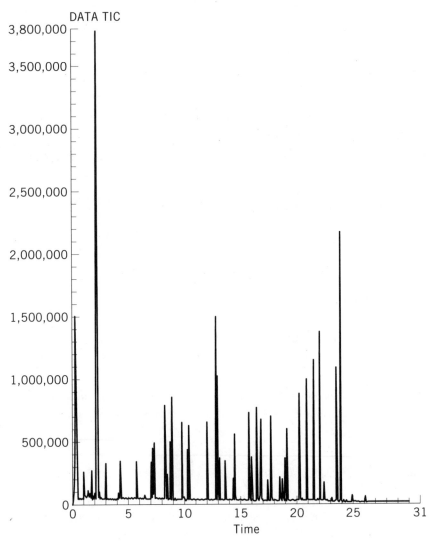

Figure 1.5 Total-ion chromatogram.

The actual data output content is much more complex. If the mass spectrometer is in the scanning (SCAN) mode, the analyzer voltage is being changed continuously and repeatedly over a selected mass range. Different ion masses are reaching and being detected by the detector. Information is coming out each moment on the exact position of the analyzer.

After calibration and in combination with the ion concentration information, this provides the molecular mass and amounts of each ion formed. After these data are computer massaged, we receive a three-dimensional block of data whose coordinates are elapsed time, molecular (m/z), and ion concentration (Figure 1.6).

Viewing this block of data on a two-dimensional display such as a integrator or cathode-ray tube (CRT) while trying to extract meaningful information is nearly impossible. A three-dimensional projection can be made but is not particularly useful for extracting information. It does provide a topological view of the data, which is useful for finding trends in the data set.

If we select a data cut at a single molecular mass, we can produce a single-ion chromatogram (SIC) similar to a UV detector tracing at a single wavelength (Figure 1.7).

The series of peaks produced represent the concentration of ions of the selected molecular mass present throughout the chromatographic run. Compounds that do not form an ion with this mass will not be present in the chromatogram. Comparison with the TIC shows a much simplified chromatogram, but all peaks in the SIC are present in the TIC.

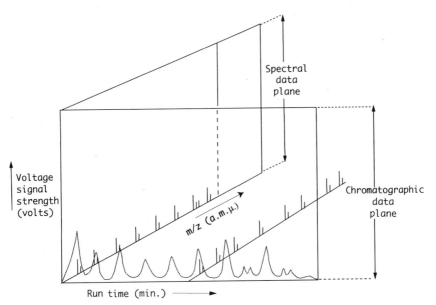

Figure 1.6   Three-dimensional GC/MS data block.

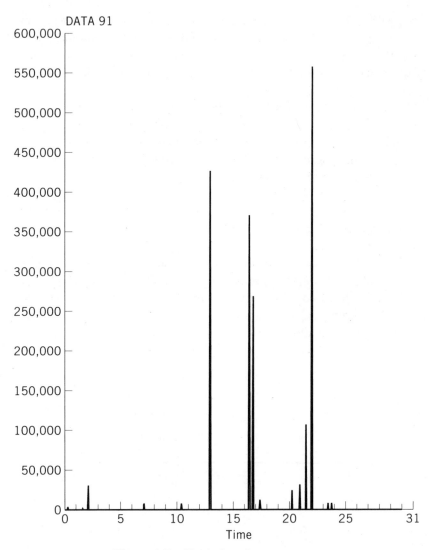

Figure 1.7   Single-ion chromatogram.

An SIC can also be produced by running the GC/MS in a fixed mass mode in which the analyzer is parked at a given molecular mass position throughout the chromatographic run. This single-ion monitoring (SIM) mode has an additional advantages. Because the analyzer is continuously analyzing for only a single ion, the ion yield is much higher and detection limits for this ion are much lower. The mass spectrometer becomes a much more sensitive detector, but only for compounds producing this mass fragment. Any other compounds will be missed. A good detector for trees instead of forests—for trace analysis of minor contaminants.

Going back to our three-dimensional block of data in Figure 1.6, we can select a data cut at a given time point that will provide us with a display of molecular mass versus ion concentration called a mass fragment spectra or simply a mass spectra (Figure 1.8).

Generally, these data are not displayed as an ion continuum. The ion masses around a unitary mass are summed within a window and displayed as a bar graph with 1-amu increments on the m/z mass axis, as shown in Figure 1.8.

The mass spectra of a resolved compound is a record of the fragmentation pattern of this compound under a given set of experimental conditions. It is characteristic of that compound and can be used to definitively identify the chemical nature of that compound. In the same or a different instrument under the same conditions, this compound will always give the same fragments in the same ion concentration ratios. Libraries of compound fragmentation patterns can be created and searched to identify compounds by comparison to known fragmentations. Further decomposition of isolated fragments can be studied in triple quadrupole GC/MS/MS systems to identify fragmentation pathways useful in determining structures of unknown compounds.

There is a lot of arm waving involved with the statement "under a given set of experimental conditions." Different ionization methods and voltages will affect the fragmentation ions produced. Under certain conditions only a single major ion is produced, the molecular ion. The original molecule loses an electron to form this ion radical, whose mass is equal to the molecular weight of the compound, a very useful number to have in identifying compounds.

Changes in the geometry, calibration, cleanliness, and detector age of the mass spectrometer can all produce variations in the fragmentation pattern and especially in the ion concentration ratios. Variations in the chromatographic conditions can lead to overlapping peaks and can change the fragmentation pattern. Learning and controlling these will convert GC/MS from a science to an art. All of this has led to a proliferation of instrument types and calibration standards to try and tame the variables.

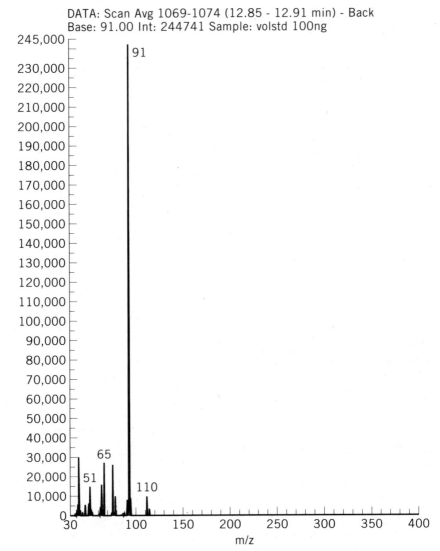

Figure 1.8 Mass fragment spectra (MASS SPECTRA).

1.4. SYSTEMS AND COSTS

Instrument system costs are not widely advertised by manufacturers unless you work for the federal government and are buying off a Government Service Administration price list. To come up with even ballpark figures, I have talked to former customers who have recently purchased systems and to manufacturers at recent technical meetings. The numbers in Table 1.1 represent an educated guess at 1997 system pricing.

In the past, systems could be divided into two basic types: floor standing systems designed for the mass spectrometry research laboratory and desktop systems designed for both commercial analytical laboratories and university analytical chemistry laboratories. A new product niche opened in the last two years. These systems are simpler, easier to maintain and calibrate, and are aimed at quality control and analytical testing laboratories. They are advertised at a third of the price of desktop systems of four years ago. The jury is still out on these, but some of their manufacturers have good pedigrees and track records.

I have included pricing estimates on triple-quadrupole MS/MS systems and on research and desktop ion trap GC/MS systems for comparison to the quadrupoles and because many users consider these to be the analytical systems of the future. The ion traps seem to be simpler, more sensitive, ideal systems for MS/MS studies. If the future is truly toward smaller, more compact systems, the ion trap may lead the way.

Overall, there is a trend toward lower pricing and ease of operation. This will make systems more available to the average research investigator and commercial laboratory.

There is a growing market for older GC/MS systems because of price and the availability of upgraded data systems, both from the manufacturer and from third-party sources. It is true that the data system is usually the worst part of the older system, computer technological advances having left them in the dust. They are difficult to learn, hard to use, and very difficult to connect into data networks since their data formats are obsolete or on the verge of becoming obsolete.

Pumping and analyzer sections almost always work. Detectors and data systems can generally be replaced if necessary. Once retrofitted, these systems usually perform like champs.

However, there are systems out there that were never very good and no amount of retrofitting will improve: systems without butterfly valves in the pumping system that dump pump oil into the analyzer in case of power failures, systems whose manufacturers have disappeared into the night or one-of-a-kind systems, with no two systems having the same control inputs or detector outputs. I have demonstrated replacement data systems on all of these. Let the buyer beware!

When retrofits work, older systems are often great buys. One customer who purchased a hardly used GC/MS system from a hospital for $25,000, added a modern data/control system for $22,500, and ended up with a

TABLE 1.1 Estimated GC/MS System Prices

MS Type	GC System	MS Only	GC/MS	AS/GC/MS Data
Quadrupole	Production	N/A	N/A	$36,800[a]
	Benchtop	$72,000	$80,00	$90,000
	Research	$90,000	$96,000	$111,000
	Triple quadrupoles	N/A	N/A	$250,000–$500,000
	Used quadrupole	>$3,000	>$5,000	$3,000–$50,000
Ion trap	Ion detector	$65,000	$72,000	$82,000
	Research MS/MS	N/A	N/A	$450,000

[a]No autosampler.

state-of-the-art system for under $50,000. A production facility just getting started bought 12-year-old systems for $3000 each, modernized the data system, networked the systems, and ran them day and night until it could afford to replace them with 20 newer systems. Bare systems were purchased, without a processing and control computer, and the data/control systems moved to each new instrument as they were purchased. Operator retraining was neglible as well as system switchover time.

The key to buying an older system is to buy one made by a company that was successful when the system was sold or that is still successful. Talk to people who have used or are still using the instrument. Find out what they think about it—its strengths and limitations.

1.5. COMPETITIVE ANALYTICAL SYSTEMS

What other analytical systems need to be considered when selecting an instrument to use in your research? To give you an idea of the size of the worldwide chromatography market in 1996, high-performance liquid chromatography (HPLC) sales led with $1.5 billion (6–8% annual growth), GC sales were $1 billion (3–4% annual growth), and capillary zone electrophoresis (CZE) sales were only $80 million (10% annual growth). Few of these systems were purchased with mass spectrometers attached, but this number will increase as mass selective detector (MSD) prices fall over the coming years.

If you need the definitive identification provided by the GC/MS system, there are few competitive systems and none at the same relatively mature state of development. On the horizon are a few contenders for the crown. One has a fairly broad application potential; others fit better in specific analytical niches.

1.5.1 Liquid Chromatography/Mass Spectrometry (LC/MS)

The high-performance liquid chromatograph connected to the mass spectrometer offers the best potential as a general MS instrument for the laboratory. These LC/MS systems aimed at the production as well as the general research laboratory are just starting to appear on the market. They claim major improvements in ease of maintenance and operator training, calibration stability, source flexibility, and system pricing (Figure 1.9).

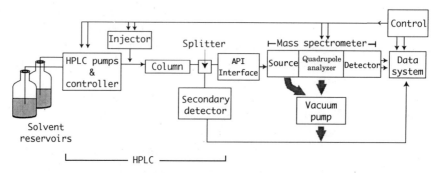

Figure 1.9   LC/MS system diagram.

Chromatographically HPLC offers flexibility in media and in isocratic and solvent gradient separation technology. Almost anything that can be dissolved can be separated, generally without much sample preparation or derivatization. Large molecules such as proteins and restriction fragments can be separated and analyzed using electrospray techniques. Limitations to using the technique are primarily due to the recent introduction of commercial LC/MS systems. Existing analytical techniques and calibration standards are just appearing, and few have been approved by regulatory agencies. Price and reliability are still considerations for general laboratory applications. Existing spectral libraries may require modification and will need additional compounds added to them.

1.5.2. Capillary Zone Electrophoresis/Mass Spectrometry (CZE/MS)

Another research tool of growing popularity, CZE interfaced with a mass spectrometer offers a powerful, but limited means for analytical separations (Figure 1.10).

Capillary zone electrophoresis uses electromotive force to separate charged molecules in a capillary column filled with buffer or buffer-containing gel. A very strong electrical voltage potential is applied across the ends of the column. Ionized compounds move toward the electrode with the opposite charge at a rate dependent on their size and charge strength. It is designed to work with very small amounts of material and delivers a very concentrated compound disk to the mass spectrometer interface.

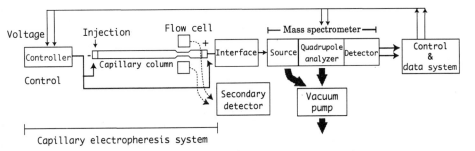

Figure 1.10   CZE/MS system diagram.

Very high efficiency separation can be achieved. It has proved very useful for analyzing multiply charged molecules such as proteins and DNA restriction fragments when combined with an electrospray MS interface. Limitations for general application have been injector design problems, necessity to work with very high voltages and high buffer concentrations, and problems eluting sample into the interface. Current system cost, high levels of maintenance, and calibration stability problems have prevented this technique from wider application, but these appear to be coming under better control. Like LC/MS there are few approved methods for production applications.

1.5.3. Supercritical Fluid Chromatography/Mass Spectrometry (SFC/MS)

Widely considered to be only a laboratory curiosity, SFC/MS has been adopted by a major GC/MS manufacturer and may be developed into a useful environment analysis tool. One of its most attractive features is its use in combination with supercritical fluid sample extraction in automated sample preparation and analysis.

In SFC/MS gases such as carbon dioxide can be used in their supercritical fluid state as a mobile phase for separation of injected material on a normal phase HPLC column. Equipment from the injector to the detector interface must be operated under the pressures needed to maintain the gas in its supercritical fluid state (Figure 1.11).

The major advantage to the technique is that the mobile phase is dispersed simply by reducing pressure. Except for minor carbon dioxide

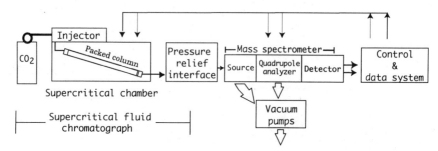

Figure 1.11   SFC/MS system diagram.

contamination, there is almost no solvent background in the mass spectrometer operation. Limitations are the requirements for high-pressure chromatographic separation, limited mobile phase selection, and the lack of availability of commercial equipment and methodology. The latter two problems may quickly disappear. The equipment currently available is too expensive and unreliable for day-to-day analysis. Again, this may change quickly.

CHAPTER 2

SAMPLE PREPARATION
AND INTRODUCTION

The mass spectrometer is designed to analyze only very clean materials. Even solvent can interfere with fragmentation pattern identification. We attach a gas chromatograph to separate materials and, obviously, it can clean samples. But separation and sample cleanup of real-world sample exceed the capacity of the device.

Before introducing the sample into the gas chromatograph, some form of sample preparation is needed. If we can combine the sample preparation and introduction into a single, automated apparatus, we have achieved our purpose. In this chapter we will look at methods for direct injection into the mass spectrometer, gas chromatography injection techniques, and extraction methods for freeing our compounds of interest from various environmental matrices.

2.1. DIRECT SAMPLE INJECTION INTO THE MASS SPECTROMETER

Many GC/MS systems have a port for a direct insertion probe (DIP) on which a sample may be inserted directly into the mass spectrometer source for ionization and analysis. The sample can be introduced as a drop of liquid, a solid, a film dissolved in a volatile solvent, or an emulsion or suspension in an activating compound for fast atom bombardment (FAB).

2.1.1. Direct Insertion Probes and Fast Atom Bombardment

The DIP port is a vacuum lock through which the probe can be inserted without disturbing the analyzer vacuum. The probe itself is a metal tube, usually equipped with an electrical heating unit, which ends in a slanting sample face that is inserted into the ionizing source beam. The sample is volatilized by the vacuum system or by programmed heating of the probe heating element (Figure 2.1).

The FAB technique is used for ionizing nonvolatile solids. The sample is ground into an emulsion in a viscous liquid such as glycerine. A drop of this emulsion is placed on the probe face and inserted into the source through the port. The sample is bombarded with heavy ions, often cesium, from a special ionization source. The droplet absorbs the collision energy, explodes, and throws some of the sample into the source cavity, where it is ionized and swept into the analyzer. If this sounds like a messy procedure, it is. Fast atom bombardment sources require frequent cleaning. But, FAB's usefulness for nonvolatile samples is great enough to make it a very popular technique.

The literature has reported attempts at combining GC separation with FAB in a technique called flow FAB. The GC effluent is eluted into a separator interface to remove part of the carrier gas and to introduce a FAB solvent. The solution or suspension is then sprayed into the evacuated FAB ionization source. The technique has also been used in time-

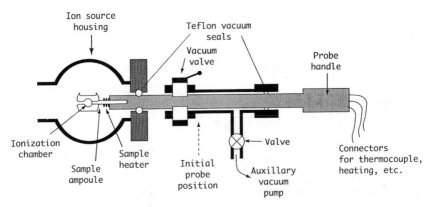

Figure 2.1 Direct insertion probes.

of-flight mass spectrometers to introduce absorbing dyes which are excited with laser sources to aid in sample ionization.

2.1.2. Head Space Analyzers

Another direct injection technique used in GC/MS laboratories is head space analysis. Sample is introduced into an evacuated chamber and then sealed. The sample can then be vented directly into the evacuated mass spectrometer source where each component can be analyzed using SIM mode. A distinctive mass ion is chosen for each component of interest, and these ions are monitored sequentially using step scan analysis. I have seen this technique used for monitoring automobile exhaust gas ratios over days, weeks, or months.

Alternatively, a head space analyzer can be set up to feed a gas chromatograph. The sample is introduced into an evacuated chamber through a vacuum port and the volatilized sample components are swept into the gas chromatograph with carrier gas where they concentrate on the column head before separation and elution.

2.2. SAMPLE PURIFICATION

Nature has a habit of creating complex mixtures. To analyze these, we turn to techniques that are compatible with the constraints of the mass spectrometer. Solvent removal and sample concentration are two of the major problems that had to be solved before the mass spectrometer could be connected to HPLC system effluents as well as for samples purified for GC/MS analysis.

The problem with purification using extractions is difficulty of getting rid of the extracting solvent before introducing the sample into the mass spectrometer. The mass spectrometer is an excellent analyzer for trace amounts of unremoved solvents. For instance, if you extract a soil sample suspended in acidified water with methylene chloride and then inject it into a nonpolar GC column, you will be dealing with a severe methylene chloride peak contamination problem if you do multiple injections. If you put the extract in a nitrogen tube dryer, evaporate the methylene chloride, and take the sample up in methanol (assuming it will redissolve in methanol) before injecting, you will still have a methylene chloride contamination problem. But, now, you will have to split off the eluted methanol or it will get into the mass spectrometer source.

Both techniques are used, but it is best to avoid solvent additions whenever possible. A step in the right direction is to use preinjection cartridge columns for extraction and concentration. Organics in aqueous solution can be partitioned on to these columns with the stationary phase acting in place of an organic solvent. The retained material can be eluted with a small amount of methanol or acetonitrile for injection into the gas chromatograph.

Use of polar cartridges in a supercritical fluid apparatus would allow extraction with supercritical carbon dioxide. With pressure reduction directly in a carrier gas stream, most compounds could be swept directly onto the GC column. "Most" is an important qualifier since many compounds would precipitate on pressure reduction and only be volatilized by heating the injector port, if at all. This technique is being evaluated for automated extraction of environmental sample for direct injection into GC/MS systems.

The next step is to purge the volatiles from aqueous samples with carrier gas directly into the GC injector port. This introduces a lot of water into the injector and suggests the next improvement, the purge-and-trap apparatus (Figure 2.2).

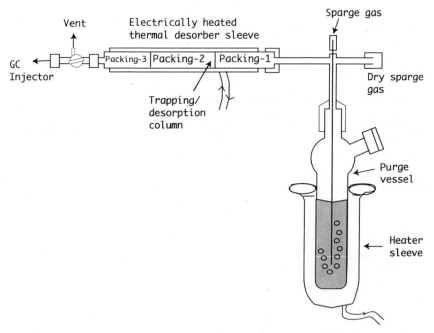

Figure 2.2   Purge-and-trap injection system.

This apparatus is made of a purge tube, which may be heated with a sleeve, containing the aqueous sample and a gas purge line. The volatiles are swept by purge gas into a packed column in which they are trapped until the end of the purge period. Unretained purge gas is vented from the system. For elution into the GC injector, carrier gas is swept in the reverse direction into the trap, which is heated to elute the trapped sample directly onto the GC column.

2.3. MANUAL GC INJECTION

Once we have a sample in a syringe, we need a way of injecting it onto the gas chromatography column. The split/splitless injector has become the model for GC injection (Figure 2.3).

It uses a self-sealing, replaceable septum for syringe injection, a connection for carrier gas to sweep sample into the injector body, an automated valve for sample diversion, a heated throat with a removable throat liner, and a seal fitting it to the top of the capillary column. Septumless syringe ports that use a spring-sealed Teflon surface have been recently introduced. It is important to use a specially designed blunt syringe needle with these ports. A pointed needle will score the Teflon surface and cause leaking around the syringe needle.

Sample is vaporized in the injector throat. The split valve is used to control the amount of sample allowed to enter the column. This is used primarily to prevent overloading the column. Since sample discrimination can occur during volatilization and splitting, a variety of throat liners are available that provide variations in surface area and composition to control these changes. The simplest throat liner is a plug of glass wool, but a variety of borosilicate glass and silica restricted tubes with constrictions are available. Specific throat liners and split times are often specified in methods. Liners must be cleaned daily to remove nonvolatile components from the injection.

Many systems are being equipped with injectors that allow direct liquid sample injection onto the column head. This is done primarily to prevent thermal decomposition of the sample in the heated injector throat. It also avoids sample discriminations and pressure variations associated with volatilization and split valving.

The obvious problem with on-column injections is that nonvolatile materials are introduced onto the head of the column and will accumulate

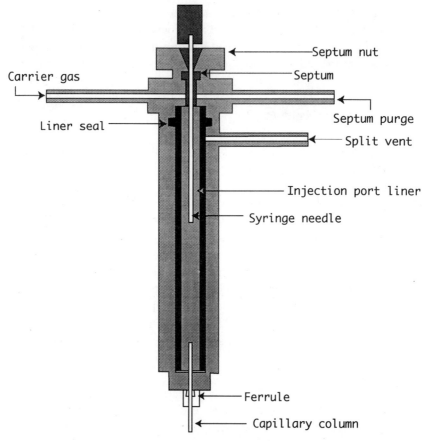

Figure 2.3 Split/splitless GC injector.

over a period of time just as they do in an injector's throat liner. There is no good way to remove them from the column; even long period bake-outs will not remove many of these compounds. They can be ignored, but many will result in longer term bleeds that can interfere with mass spectrometer operation. The simplest solution to the problem is to cut a few centimeters off the column inlet periodically, since it will have little effect on a 25–150-m column.

Carrier gas pressure is the primary driving force in controlling eluting time from the capillary column. Variation in the injector pressure will lead to variations in chromatographic retention times. Automated electronic pressure control (EPC) has been introduced to control this variable. It also offers potential as a gradient control variable similar to solvent programming in HPLC.

By reducing the pressure at specific points in the chromatographic run, compressed areas of the separation can be allowed further interaction with the column to improve resolution. Widely separated peaks can be eluted more rapidly by increasing the pressure, decreasing total run time. Method development using this technique has just begun to appear in the literature. The jury is still out on its effectiveness in improving separation while reducing run time.

2.4. AUTOMATED GC/MS INJECTION

The simplest form of injection automation is the injection trigger. This usually consists of a lever or a ring around the septum that is depressed when the sample is injected. It acts as a contact closure to send a signal to the oven to start a temperature program, to the EPC valve to run a pressure program, or to the data system to start the mass spectrometer scan and to begin acquiring data.

If you are analyzing multiple samples per day, running night and day, you will need to automate your sample injection. Autosamplers are robotic arms that pull a sample from a specific sample vial in a carousel and inject it into the injector body connected to the capillary column (Figure 2.4).

Each system provides some way of washing the injection needle between injections. They also may have provisions in their programming to allow repeated injections from the same vial or periodic returns to a series of standard vials to make calibration check injections. They may provide for positive vial identification by reading a bar code label on the vial. This number allows confirmation of the identity of the sample that has just been analyzed by the mass spectrometer. Some autosamplers provide sample carousel cooling to prevent sample degradation during standing in solution in a long series of runs.

Sample vials are usually sealed with a septum that is penetrated by the injection needle after the vial is positioned by the arm. Sample either is placed in the needle by suction from the needle line or is pushed in by hydraulic pressure on the sample surface. Most autosampler needles are filled by suction: either by a syringe connected to the needle line or by a mild vacuum on an automated valved line.

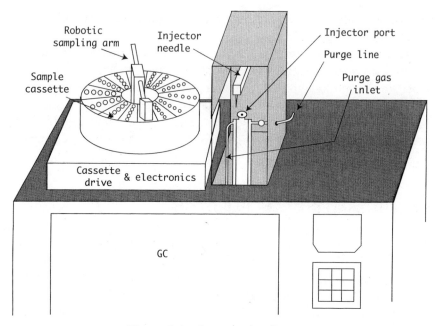

Figure 2.4   Autosampler diagram.

Once filled, the needle is swung into position over the injector septum. Injection is done by mechanically inserting the needle through the septum, then reversing the syringe drive or valving over to a low-pressure line. Once injection is complete, the needle is removed from the septum, rinsed, possibly air dried, and positioned for the next sample.

The purge-and-trap apparatus can also be automated. A number of automated systems on the market can be loaded, programmed, and left to run unattended. Sample heating, purge gas flow, carrier gas flow, and trap exhaust valves are all automated from a programmable microprocessor based controller. Trap heating temperature and heating time are also under programmer control.

A purge-and-trap system currently on the market will automate up to 16 purge tubes for liquid or solid samples. Another system uses only a single tube for liquid sample purge but feeds samples into the purge tube from an autosampler. Obviously, some provision must be made for purge tube washout between samples to avoid sample cross contamination. Both forms of automated purge-and-trap systems are specified for use in Environment Protection Agency (EPA) procedures.

CHAPTER 3

THE GAS CHROMATOGRAPH

A gas chromatograph is a programmable oven designed to run GC columns. It has a microprocessor-based controller whose major purpose is to provide temperature gradient programming. A secondary function of the controller is to provide automated actuation of switching valves.

3.1. THE GC OVEN AND TEMPERATURE CONTROL

Oven temperature controllers usually have at least five programmable linear ramping segments. In creating a temperature profile program, linear ramps with differing slope rates are linked to achieve separation at different points in the separation. Purge gas flow rate and auxiliary valve switching can be set and changed at time points along the temperature ramps. The oven program is usually started with an injection signal but can also be started from the keyboard of the gas chromatograph or from the mass spectrometer's computer control panel. The ideal is to arrange the system so that one run signal starts all of the run components.

Internally the gas chromatograph is made of three compartments (Figure 1.2). The electronic enclosure provides space for the microprocessor boards, with the display and keyboard on its exterior face. An unheated area holds the injector port head, injector trigger, purge gas lines, valves, and cabling. The large, insulated cube of the oven, with a door making up its front face, holds the injector body (the column) and provides an exit port for the mass spectrometer interface connected to the end of the column. Secondary detectors are connected via a T-splitter line to the outlet of the capillary column with the detector body and flow cell either on top of the gas chromatograph or in the unheated cable/injector area.

To shorten chromatographic run times, automated external cooling of the GC oven for temperature reequilibration may be provided. This may be as simple as a mechanism to open the door of the gas chromatograph until the temperature drops. Or there may be provisions for adiabatic cooling using compressed carbon dioxide gas, a technique called cryoblasting.

3.2. SELECTING GC COLUMNS

The typical gas chromatographic column used for GC/MS is a 25–150-m coiled capillary tube with an internal diameter of 0.25–0.75 mm. Drawn from either glass or silica, it has an activated surface and an internal coating of a viscous liquid such as Carbowax that acts as the stationary phase of the gas–liquid separation. Figure 3.1 illustrates the technique used in drawing and placing a protective epoxy coating on a typical capillary column.

Since coated columns have a slow bleed of the stationary phase, new columns have been created with the stationary phase crosslinked and chemically bonded to the silica wall. These columns are more stable, give more reproducible results, and have longer working lives.

A wide variety of stationary phases with different chemical composition are available. The most common film or bonded phase is nonpolar material such as methyl silicone. This packing is stable, has a high capacity, and provides a separation that parallels the compounds' boiling points. Low boilers come off first, high boilers last. This column is also described as a carbon number column since the more carbons in the compound's structure, the later it comes off the column. Other columns, such as Carbowax, phenylcyanopropyl-, and trifluoropropyl-, either are more polar, showing an affinity for hydrogen-bonded compounds, or show an affinity for compounds with functional groups or for compounds with a high dipole moment. These columns can be used to optimize the separation of compounds that are not resolved using a simple boiling point column. Up-to-date catalogs from column suppliers usually give guidelines in selecting a column for a specific separation.

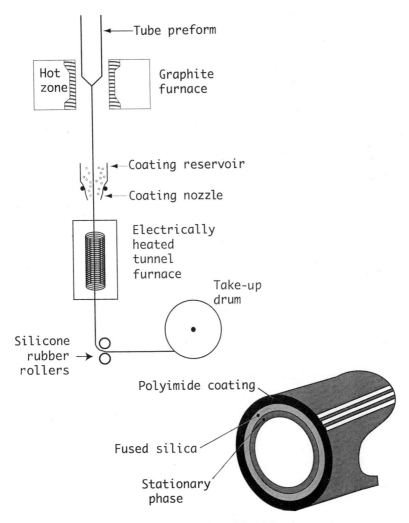

Figure 3.1   Preparation of the GC column.

For analysis under a method specified by a government regulatory agency, there may be little choice about the nature of the column selected for the work. Both coated and bonded columns are used in commercial analysis laboratories, but bonded columns are the choice for new method development.

The final step in column preparation is to cover the outside of the column with a polyimide or metallic coating to provide protection against shattering (Figure 3.1). Column coating and bonding are an art best left to the expert.

There is always some variations between columns when running standards. If column to column reproducibility is critical, select a column from a reliable manufacturer and stay with it. Check the quality control provided by a new column against your own standards when the column comes in. Don't jump from manufacturer to manufacturer because of price. Your time and the results you put out are too important to risk them to save a dollar on a column.

For preparative GC, you may find a need for an open-tube, coated column. The only time to select anything but a bonded-phase, capillary column for MS is if the method you are trying to duplicate specifies one. Even then, first try to find another method that does the same thing on a capillary or a bonded-phase column. The efficiency and stability gained will always be worth the effort and time.

3.3. SEPARATION PARAMETERS AND RESOLUTION

Optimum chromatographic separation is achieved with baseline resolution between all adjacent peaks in a reasonable run time. To achieve this type of separation, it is necessary to understand the variables that can affect a separation.

In Figure 3.2, we have a two-compound chromatograph run at high strip chart speed so we can measure system parameters that can be used to define the separation.

Two parameters are measured, the retention time t_r and the peak width w of each peak. We also need to measure or know the void volume time of the column, $t_{r,0}$, the retention time of an unretained solvent.

From these data we can calculate a retention factor k, (also called the capacity factor), a separation factor α, and an efficiency factor n. Finally, using these values, we can produce a resolution equation combining all of these factors. The values of these numbers may be important to the theoretician, but I find them useful mainly for predicting the changes that can be made in a separation or as a diagnostic aid in following separation changes over time.

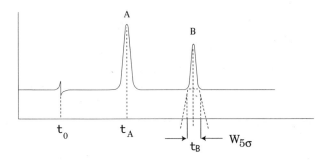

Retention: factor
$$K_A = \dfrac{t_A - t_0}{t_0}$$

Height equivalent: to a theoretical plane
$$h = \dfrac{L}{n}$$

Separation: factor
$$\alpha = \dfrac{t_B - t_0}{t_A - t_0} = \dfrac{k_B}{k_A}$$

Resolution equation:

Efficiency: factor
$$n = 5.4 \left(\dfrac{t_B}{w_{\frac{1}{2}}h}\right)^2 \quad \text{or} \quad 16\left(\dfrac{t_B}{W_{5\sigma}}\right)^2$$

$$R_5 = \dfrac{1}{4}\left(\dfrac{\alpha-1}{\alpha}\right)\left(\sqrt{n}\right)\left(\dfrac{k}{k+1}\right)$$

Figure 3.2   Gas chromatograph parameters.

Looking at the chromatogram in Figure 3.2, we can see that if $t_{r,0}$ is set to 1, $t_{r,A} \sim 0.5$, $t_{r,B} \sim 1$, and $\alpha \sim 2$. Since we have baseline resolution between the two peaks, we have a usable separation. However, we may not have an optimum separation if the run time is too long. It should be noted here that all peaks do not have to have baseline resolution in a usable separation. Someone has to make a decision on how perfect a separation must be to be used for a particular analysis. Reproducibility is often more important than separation perfection in the real world.

The efficiency factor n tells how sharp the peaks are and how much overlap is occurring between adjacent peaks. The sharper the peaks, the closer I can run them together, the faster I can separate them, and the shorter will be my overall run time.

To measure efficiency, we must measure peak width at a given retention time. The longer a peak is on the column, the wider will be its peak width due to diffusion. The most common width measured in GC is the width at half peak height because GC peaks are very symmetrical. If your peaks tail on the back side, as they do in HPLC, you are better off using the second efficiency calculation, which measure the peak width, called the 5-sigma width, at approximately $\frac{1}{10}$ the peak height. It is much more sensitive to column changes and contamination than the half peak height measurement. It is measured by extending each side of the peak slope until it meets the baseline; then the baseline segment formed is measured. The two methods of efficiency calculation are identical for symmetrical peaks.

Efficiency is reported in theoretical plates per meter. The larger the plate count, the higher the efficiency, and the sharper should be the separation peaks. A related value is the height equivalent to a theoretical plate, h, or the column length divided by the efficiency. In h, higher efficiency leads to smaller numbers. Plots of h versus flow velocity are used to select optimum flow rate for columns.

Finally, efficiency, retention, and separation factors are combined into the resolution equation. The resolution equation shows that the retention factor portion of the equation is a convergent term. It has a significant initial effect on resolution, but it falls off as the retention factor increases. The efficiency factor portion is a square root function. Changes to efficiency do not produce a linear response in resolution. Changes in the separation factor portion of the equation have a nearly linear effect on resolution. So what are the variables that affect retention, separation, and efficiency? These will be discussed next.

3.4. GC CONTROL VARIABLES

Temperature is the major control variable, followed by carrier gas pressure, which controls gas velocity. It would be preferable if each variable changed only one resolution factor at a time. Some do, but many exhibit complex effects on more than one factor. Only a few variables can be used as a control variable once the column and carrier gas has been selected.

The variables affecting separation are as follows (the first five are selected before starting the run):

1. *Stationary Phase Chemistry.* Column chemistry changes produce α effects that lead to switching of the relative-peak positions from column type to column type. On one column the peaks may elute a, b, and c; on another they may elute a, c, and b or two peaks may coelute. Traditional GC column packings such as Carbowax separate compounds pri-

marily by carbon number: The more carbons in the molecule, the longer it is retained. New bonded-phase columns with altered surface chemistry offer great potential for taking advantage of alpha changes to separate unresolved compounds. A number of new column types have appeared in the last few years, but they have only slowly been adopted for separations. If you cannot make a separation, try changing the type of column you are using. For instance, supports containing aromatic compounds should have an affinity for double bonds and aromatic compounds.

2. *Stationary Phase Thickness.* Thicker coats increase retention time k because the sample has more opportunity to interact with the stationary phase. Thick phase columns are used for analysis of light components, thin phases for heavier components. With heavily crosslinked supports transfer through the support thickness is inhibited, and this can be overcome with increases in gas pressure and temperature. Be aware that nonbonded, thick-coat columns are susceptible to dramatic column failure on heating or shock as the column support separates from the wall and beads up.

3. *Column Internal Diameter.* Decreasing the column diameter increases both efficiency n and retention k. Less material is channeling down the column center and the ratio of gas to liquid phase favors better interaction with the column. Retention times are increases as well as total run times.

4. *Column Length.* The length of column has an effect on efficiency n, but the resolution equation tells us that the change is related to the square root of the length change. The longer the column, the more interaction occurs and the greater is the efficiency of the separation. Resisting this is the turbulent diffusion of the separated samples, which leads to band broadening and decreased efficiency. Shortening a column to remove nonvolatile materials or plugs will decrease efficiency, but not so you will notice it right away unless you start hacking off big chunks. Do it in moderation; plugs and nonvolatile compounds are trapped in the first few centimeters.

5. *Carrier Gas Chemistry.* The chemical nature of the carrier gas can have a dramatic effect on the efficiency n of column operation. Because of its relatively high viscosity, nitrogen is a poor carrier gas with a low range of usable gas velocity efficiency that drops off rapidly at high flow. Helium and hydrogen are both better choices, with hydrogen the gas of choice. Because of its explosive nature, proper venting of hydrogen is important.

6. *Carrier Gas Pressure.* Increasing pressure increases the retention time k of sample in the stationary phase. The sample has a longer period to interact with the column and to improve separation. In systems offering programmable electronic pressure control this variable offers real potential for gradient control in methods development.

7. *Temperature.* Temperature is the major control variable used in gas chromatography. Elevated temperature decreases retention time $k,$ but it also can lead to altered separation effects α. Peak positions do not always maintain their relative position as temperature is increased. This can be useful when the effect causes peak changes in the correct direction, but the effect is difficult to predict. Because the effect is not instantaneous, there is a lag time that varies with the oven design. This leads to some variations in methods when running samples on different manufacturers' equipment. Electronic pressure control offers to be a simple retention factor variable with instantaneous response.

3.5. DERIVATIVES

To be successfully analyzed by GC, a compound must be volatile. But what do we do if it is not? One of the techniques used is to derivatize the compound, which often increases its solubility. But, why would adding mass to a molecule increase its volatility?

The clue to understanding what makes this work is in the kinds of functional groups that are derivatized. Compounds such as BSA (bis-trimethylsilylacetamide) or BSTFA (bis-trimethylsilyltrifluoroacetamide) place trimethylsilyl or trifluoromethylsilyl groups on active hydrogen sites in amines, alcohols, or carboxylic acid groups. Alkyl ester derivatives are formed from carboxylic acid; oximes are made from ketones and aldehydes. When treated, all of these groups have two things in common: They form hydrogen bonds and they aggregate. Hydrogen-bonding interaction reduces the volatility. Derivatives that prevent hydrogen bonding or remove hydrogen-bonding functional group increase a compound's volatility.

Crude samples that must be derivatized before analysis generally have to be extracted into an organic solvent and dried before being reacted. Derivatizing agents will react with the hydrogens in water as readily as with the target compounds. Catalysts such as trimethylchlorosilane or reagents such as pyrimidine must sometimes be added to complete the reaction. All reactants added to the mixture are potential contaminants for the mass spectral analysis and must be removed, either by GC or by preinjection extractions. If possible, avoid derivatization; it simplifies the separation.

CHAPTER 4

THE MASS SPECTROMETER

The basic components of the mass spectrometer are the pumping system, the interface to the gas chromatograph, the ionization chamber and electron source, the focusing lens, the quadrupole analyzer, the detector, and the data/control system. Pumping systems providing high vacuum ($<10^{-5}$ torr) are critical to the operation of the mass spectrometer. Electrons and ionized compounds cannot exist long enough to reach the detector if they suffer collisions with air molecules in the analyzer.

4.1. VACUUM PUMPS

Vacuums for mass spectrometry are established in two stages: a forepump takes the vacuum down to 10^{-1}–10^{-3} torr, then either an oil diffusion pump or a turbomolecular pump drops the analyzer pressure to 10^{-5}–10^{-7} torr (Figure 4.1).

Vacuum is measured in either torr or pascal units. The torr, equal to the pressure of 1 mm Hg, is a commonly accepted vacuum measure in the United States. The pascal, equal to 7.5×10^{-3} torr (mm Hg), is more commonly used in Europe.

Vacuum pressures are measured by two types of gauges. The medium-level vacuum of the forepump can be measured by a thermoconductivity

A) Rotary vane vacuum pump

B) Oil diffusion pump

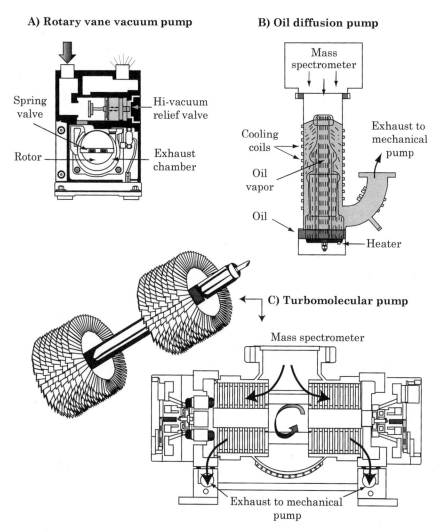

Figure 4.1 Mass spectrometer vacuum pumps.

gauge, such as a Pirani gauge. A heated wire is exposed in the vacuum line and is cooled by contact with molecules. The lower the contact rate, the lower the current draw, and the lower the vacuum. High vacuums produced by an oil diffusion or turbomolecular pump require use of a hot cathode gauge. Electrons streaming from the cathode are lost through contact with air molecules. The current produced is proportional to the concentration of air molecules present.

The mechanical roughing or forepump is an oil-sealed, rotary-vane vacuum pump commonly used as the laboratory workhorse vacuum pump. A piston on an eccentric drive shaft rotates in a compression chamber sealed by spring-loaded vanes and moves gas from the inlet side to the exhaust port. It can only reach 10^{-3} torr vacuums because of the vapor pressure of the sealing oil. Mechanical pumps typically exhibit pumping capacities of 50–150 L/min.

The oil diffusion pump sits between the inlet port of the rough pump and the outlet of the mass spectrometer. Vacuums should be below 10^{-2} torr before the diffusion pump heater is turned on. Heated oil rises up the pump chimney, jets out through circular openings at various levels, condenses on contact with the cooled walls trapping gases from the mass spectrometer, and runs down the sides, exhausting entrained gases into the roughing pump inlet. Diffusion pumps reach vacuums of 10^{-9} torr when chilled with liquid nitrogen. They can have capacities as high as 200–500 L/s, which can be important when pumping mass spectrometer sources using high levels of gases for chemical ionization or when running ion spray HPLC interfaces. Many systems with oil diffusion pumps have butterfly valves that snap shut in case of power loss to prevent contamination of the analyzer. This is an excellent feature if you are responsible for cleaning the analyzer.

The turbomolecular pump, commonly referred to as a turbo pump, is the jet engine for the mass spectrometer. It has a series of vaned blades on a shaft rotating at speeds up to 60,000 rpm between an alternate series of slotted stator places. Air is grabbed by the blades, whipped through the stator slots, and then grabbed by the next blade. Only a small amount of air is moved each time, but the number of blades and the high rotary speed rapidly move air from the analyzer chamber to the exhaust into the rough pump. Most turbos have a dual set of vanes and stators on a single shaft feeding a dual exhaust.

The turbo pumps on many desktop systems are only the size of two fists, but they can bring the analyzer pressure down to 10^{-8} torr. They do not have the pumping capacities of larger turbos, which can move 150–2500 L/s. They are used on systems having only electron impact (EI) interfaces that do not have high source pressures. The biggest advantage to the turbo pump is that it contains no oil to contaminate the analyzer. Its biggest drawback is mechanical failure, although that has been constantly improving. Work with a manufacturer that has a good trade-in program.

It is important to vent a turbo pump to the atmosphere before turning it off. The same is true of an oil diffusion pump heater. Oil vapors can be sucked into the turbo pump from the rough pump if it is left under vacuum. Some systems use all three pumps: a rough pump connected to an oil diffusion pump on the source connected to a turbo pump on the analyzer. In shutting down these systems, turn off the diffusion pump heater, allow it to cool below 100°C, vent the system, and then switch off the turbo pump. These differentially pumped systems are very important when running a chemical ionization source where there are very high source pressures. They also allow GC effluent to be run directly into the source without using a separator interface to reduce the volume of sample.

4.2. INTERFACES AND SOURCES

The interface between the gas chromatograph and the mass spectrometer is critical for system performance. It transfers sample from the gas chromatograph into the source without mixing separated bands. It can be designed as a separator to concentrate the sample about 50-fold and to reduce the source pressure by removing much of the carrier gas. It can be designed to exchange the carrier gas with a makeup gas to aid in running chemical ionization. That's the good news. The bad news is you are probably stuck with the interface that the manufacturer selected when optimizing your system.

The basic interface is a direct connection of the capillary column end into the sample inlet port on the ionized source. A differentially pumped MS system with a high-volume-transfer pumping system on the source and a separate high-vacuum system for the analyzer would be able to handle the complete GC column feed. On a mass spectrometer with a single source of high-vacuum pumping, this direct connection may supply sample too fast. The vacuum system may not be able to maintain the 10^{-4} torr vacuum needed for ionization. Too many ion-to-molecule collisions would occur to provide an ionized sample stream to the analyzer.

The next improvement would be to add a T-splitter connection with a needle valve on the mass spectrometer side. This allows a control amount of sample to be diverted to waste or a secondary detector (Figure 4.2).

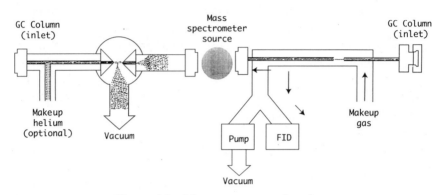

Figure 4.2   Mass spectrometer interface.

Improving on this, we can add a mechanical pump vacuum system on the exhaust of the T and add a jet separator in the capillary line from the gas chromatograph to the mass spectrometer. Sample from the capillary column expands into a separator chamber. Because of its low molecular mass, carrier gas is easily diverted into the vacuum exhaust. The high-molecular-weight samples maintain their momentum into the MS source. There is a loss of sample mass, a much higher loss of carrier gas, and a net reduction of the sample stream pressure in the source.

A number of sources have been designed for mass spectrometer sample ionization; electron impact (EI), chemical ionization (CI), fast atom bombardment (FAB), and field ionization (FI), the EI being the most common. Only the first two are commonly used in GC/MS laboratories.

The EI source (Figure 4.3) exposes the sample from the GC interface to a stream of 70-eV electrons from the filaments.

The sample molecules have an electron knocked off, or expelled, leaving behind a molecular ion with a positive charge. This ion is forced from the ionization chamber by a positively charged repeller on the back wall. The stream of ions passes through a slit, or pinhole, into a series of electrically charged focusing lenses and into the quadrupole analyzer area. The analyzer vacuum of 10^{-5}–10^{-8} torr helps move ions and prevent collisions with uncharged molecules or with each other.

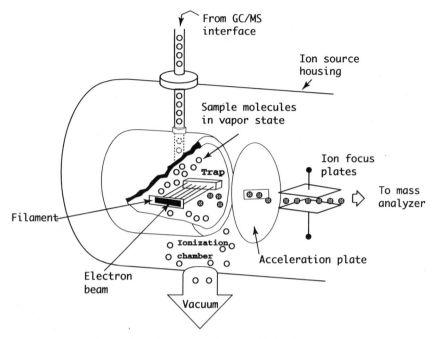

Figure 4.3   Mass spectrometer ionization sources.

The 70 eV energy of the impacting beam is high enough not only to ionize the sample molecule but also to cause many of them to fragment. The fragmentation pattern of the ions formed at a given electron energy is characteristic of the ionized molecules (Figure 4.4).

Every time a molecule of the same compound is ionized under the same conditions, it forms the same quantity and pattern of ions. This fragment pattern becomes a fingerprint that can be used to identify and quantitate the molecule being analyzed. The limitation to the technique is that under the voltage used many molecular ions first formed do not survive fragmentation. Since this molecular ion gives the molecular weight of the compound, it is sorely missed when it is absent from the EI spectrum.

This brings us to the second mass spectrometry source, the CI source. The CI source uses an ionization gas mixed with the sample stream in an enclosed ionization chamber. Gases such as methane, butane, and carbon dioxide are used to absorb the initial ionizing electron. Since the diluent

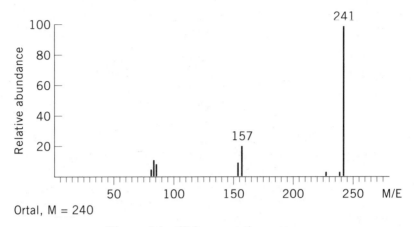

Figure 4.4 CI fragmentation pattern.

gas is present in much higher concentrations than the sample molecules, its molecules have a much higher probability of being struck by the electron stream and losing an electron. Through collision, they meet and transfer energy through a chemical process to the sample molecule, which is ionized, in turn freeing an uncharged gas molecule. The CI ionization of the sample molecule occurs at much lower energy than in EI ionization (Figure 4.5).

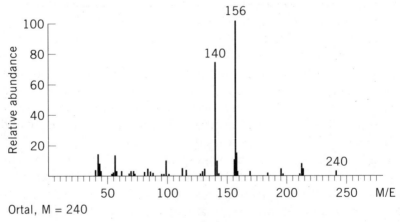

Figure 4.5 EI fragmentation pattern.

Molecular ions are retained with very little fragmentation. When analyzed in the quadrupole, the molecular ion appears as a very strong, if not the major, fragment in the mass spectrum. Since it is the largest fragment present, it can be used as quick identification of the molecular weight of the sample molecule.

Be aware, however, that sample preparation artifacts such as sulfuric acid adducts can produce compound molecular ions with masses larger than the expected molecular weight. Another problem is that some compounds do not form stable molecular ions even under CI conditions and may only exhibit a faint molecular ion fragment.

4.3. QUADRUPOLE OPERATION

Once the sample is ionized, it and its ionization fragments must be focused, propelled into the analyzer, and selected, and the number of each fragment formed must be counted in the detector.

The first step in moving the charged ion fragments into the analyzer is provided by a repeller plate at the back of the ion source equipped with a variable voltage charge of the same sign as the ionized fragments. This forces the ions through a pinhole into the higher vacuum area of the analyzer. Just past the entrance hole is a series of electrical focusing lenses (Figure 4.6).

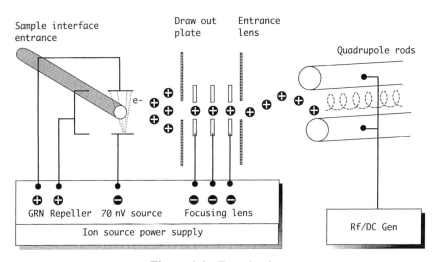

Figure 4.6 Focusing lens.

Variable voltage charges with the same charge polarity as the sample ions on these lenses squeeze the ion beam into an intense stream as it enters the quadrupole analyzer.

The quadrupole mass analyzer is the heart of the mass spectrometer. It consists of four cylindrical quartz rods clamped in a pair of ceramic collars. The exact hyperbolic spacing between diagonally opposed rods is critical for mass spectrometer operation. Rods should not be removed from the ceramic collars except by a service organization.

Both a direct current (DC) and an oscillating radio frequency (RF) signal are applied across the rods, with adjacent rods having opposite charge (Figure 4.7).

The ion stream entering the quadrupole is forced into a corkscrew, three-dimensional sine wave by the quadrupole electromagnetic field of the analyzer. The combined DC/RF field applied to the rods is swept together higher (or lower) field strength by the DC/RF generator, upsetting this standing wave for all but a single fragment mass at a given frequency. This single mass follows a stable path down the length of the analyzer and is deflected onto the surface of the detector. Any ion fragments not passed at a given DC/RF frequency follow unstable decaying paths and end up colliding with the walls of the quadrupole rods. As the DC/RF fields are swept up or down, larger or smaller masses strike the detector.

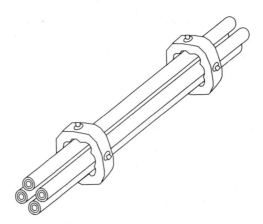

Figure 4.7   Quadrupole analyzer.

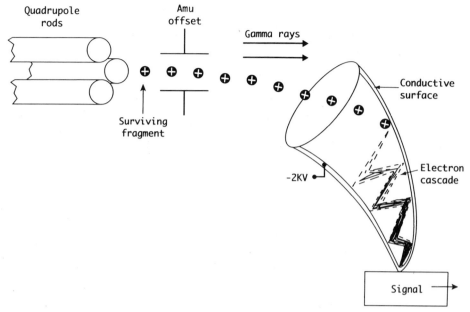

Figure 4.8   Ion detector.

4.4. THE ION DETECTOR

The fragments that pass through the analyzer strike the surface of the detector after first being deflected away from the linear path out of the analyzer by a lens called the amu offset (Figure 4.8).

Gamma particles produced in the electron ionization source are not deflected but cause false signals if allowed to directly impact the detector surface. The fragment ions striking the detector surface induce a cascade of ions in the detector body, which amplifies the single-fragment signal, sending a signal strong enough for the data system to process.

The combination of fast analyzer scanning, fast detector recovery, and high-capacity data systems allows acquisitions of about 25,000 data points per second. This means that a mass spectra run in a SCAN mode of 35–550 m/z can average 8–10 scans in 1 s. Run in a single ion (SIM) mode, the same mass spectrometer could analyze 10 m/z regions in a step scan and acquire a tremendous gain in sensitivity by averaging fifty times as many measurements at each of the 10 points.

CHAPTER 5

GETTING STARTED IN GC/MS

The purpose of this chapter is to walk through the procedure of setting up and running a GC/MS system. We will step through from an injection, produce a scanned total-ion chromatogram (TIC), extract a spectra, and do a library search.

We will simulate setting up by doing a sample extraction, equilibrating the GC oven, getting the mass spectrometer under vacuum, and programming the run. Next we will calibrate and tune the mass spectrometer for the run. We will then run a manual injection of the sample and collect the data chromatogram. Finally, we will examine the data, extract a single-ion chromatogram and a mass spectrum, and run a library search of an on-line database to identify the compound.

5.1. MODE SELECTION

Before we can start evacuating the mass spectrometer, we must make a choice of the mode of analysis we will be making. The electron ionization (EI) mode will give us fragmentation information that we can use to identify the compound. The chemical ionization (CI) mode will allow us to determine the compound's molecular weight and may give some fragmentation information. Each mode requires insertion of a different ionization source before the mass spectrometer is put under vacuum. Many desktop systems may not have the option of using a CI source. They may not have an available CI source module or sufficient pumping capacity to evacuate the source chamber against the high concentration of diluent gas. Since, in this run, we plan to do a library search on a fragmentation spectra, we will have to use the EI source. Most existing libraries are standardized on 70-eV EI ionization data from quadrupole and magnetic sector mass spectrometers.

The next mode selection we need to make is how we will scan the mass spectrometer. We can choose to do a continuous scan over a range of masses (SCAN mode) or we can do a jump scan over a discrete number of masses (SIM mode). It depends on whether we wish to look at the forest or the trees: We can choose to see a broad overview (SCAN) or a specific one with very high sensitivity (SIM). For unknown or complex mixtures we almost always run SCAN, at least for a first run. When we are looking for trace amounts of specific compounds or looking for changes in composition over a very long period of time, we would choose SIM.

Here we have chosen to run a scan from 50 to 550 amu. We select scans above 50 amu to avoid traces of water (18 amu), nitrogen (34 amu), and oxygen (36 amu) from any air residues, although we might look for these in our system tuning and performance evaluation. We also will probably be scanning from high mass to lower mass, because we get less tailing and therefore better resolution between any mass peak M and its carbon isotopic peak $M + 1$.

5.2. SETTING UP

Before making an injection, the capillary column must be connected to the injector and the mass spectrometer, the mass spectrometer must be at an acceptable vacuum, an appropriate clean throat liner must be in the injector, and the column oven must be at the equilibration temperature and programmed for the run. The sample must be prepared for injection. The mass spectrometer must be calibrated and the scan range for the run programmed before we may proceed. In most cases, once the daily autotune has been run, the only thing needed is to program the GC oven and set the mass range before making the injection.

The column we will use for the injection is a DB-5 capillary column containing a bonded phase of 5% phenyl silicone and 95% methyl silicone.

Connect the column head through the ferrule into the injector body, slide the fitting and ferrule over the column tail, and connect it to the mass spectrometer interface.

Once the source body, ionization filament, and lens are inserted into the quadrupole, we are ready to begin evacuating the mass spectrometer. In most laboratories, the mass spectrometer is usually kept under vacuum at all times unless it is down for cleaning.

To start a mass spectrometer, first turn on the mechanical oil pump and pump until a vacuum of 10^{-3}–10^{-4} torr on the ionization gauge has been reached. At that point, the high vacuum pump(s) can be turned on. For an oil diffusion pump, turn on the oil heater element and begin jacket coolant circulation. If the high-vacuum pump is a turbomolecular pump, it can be switched on at this point. It should be noted that when reversing the process and shutting down, the turbo pump should be vented before shutting off the mechanical vacuum pump to prevent oil backup into the turbo pump. When pump pressure reaches 10^{-6} torr, calibration can be initiated. Do not be in a hurry since this might take 4 h or longer.

A clean, tight system with high-capacity pumping may reaching 10^{-7} torr; if vacuum fails to go below 10^{-5} torr, start checking for leaks. When starting the evacuation, listen for a change in the sound of the rotary vane pump. If you do not hear it within 10 min, push down on the lid of the vacuum containment system to make sure the gasket has sealed. Turn on the filament and scan from 0 to 50 amu to see if there are water and air peaks. Check for leaks in the system.

Sample preparation may be as simple as dissolving a sample of mixture in a solvent. Or you may have to first extract the sample from an acidified aqueous phase, dry it, evaporate it, and derivatize it. On real unknowns add at least one internal standard, to correct for injection and retention variations, and possibly a surrogate standard, to correct for sample recovery during extraction. Also run sample blanks and extraction blanks.

If you do not know the history of the last run on the column, you may want to run a quick bakeout before setting equilibration. Turn on the purge gas, run from 50 to 300°C at 30°C/min, hold for 2 min, then step down to the equilibration temperature. Check the column specification for the maximum purge temperature for the column, especially when using a non-bonded-phase column. Overheating can cause excessive bleeding and support separation.

When the sample is ready for injection, turn on the gas chromatograph and set the equilibration temperature for the injection. This is usually around 50–120°C, but we will use 50°C for this injection. We can now program the column oven for our run. We will set a hold at 50°C for 1 min, then ramp from 50 to 320°C at 30°C/min.

Since we decided to use SCAN mode for our run, we need to select a scan range of 50–550 amu. We will work above 32 amu to avoid contamination from the air in the sample. The mass spectrometer filament needs to be protected from the slug of methanol from the injection rushing down the capillary column. If we had one, we could set auxiliary valve 1 after the GC column outlet to open from 0 to 2 min to divert solvent away from the mass spectrometer. In our case we will use a time program to simply avoid turning on the mass spectrometer filament until after the methanol bolus has passed through the source chamber.

5.3. MASS SPECTROMETER TUNING AND CALIBRATION

Before a mass spectrometer can be used to measure masses of fragmentation ions, it first must be calibrated and tuned. Calibration means adjusting the DC/RF signal frequency so that the mass axis points correspond to the expected mass fragment from a calibration compound. Tuning is done using a tuning lens to ensure that adjacent mass peaks overlap as little as possible and relative peak heights pairs have the expected ratios along the mass axis. Calibration and tuning are done so that the same compound run on different machines under the same operating conditions will always exhibit the same fragment masses in the same relative amounts. In many machines, a single tool called autotune can adjust both calibration and tuning. It usually provides an adequate calibration, but additional tuning is usually required for separation of complex mixtures.

Calibration is done with a volatile liquid called perfluoro-*t*-butylamine (PFTBA or FC43), referred to as calibration gas (Figure 5.1).

Calibration gas is placed in a vial valved off from the source. When it is needed, the valve is opened, and some of the PFTBA volatilizes into the sample chamber and is ionized. The fragmentation pattern produced has characteristic bands at 69, 131, 219, 264, 414, and 502 amu that are used to adjust the mass axis. Generally adjustments are made first on the 69 peak, which should be the largest in fragmentation pattern, and then working out toward the 502 peak, which is the smallest.

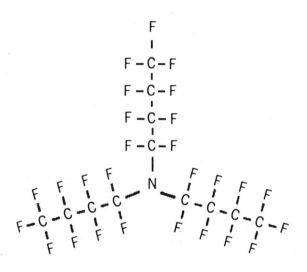

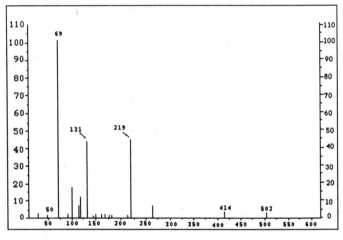

Mass	% Relative abundance
50	1.2
69	100
131	45
219	45
414	1.9
502	0.9

Figure 5.1 PFTBA calibration compound.

For our first run, we will turn on the calibration gas, then run autotune. When autotune is done, we will inspect the spectrum produced. Mass 69 should be the major, or base, peak; 131 and 219 should be about equal and a little over one-half as large as the 69 mass; and 502 would be there at around 2–5% of the 69 peak. We will discuss variables that control peak heights, lens adjustments to vary them, and other tuning compounds in Chapter 7.

5.4. SAMPLE INJECTION AND CHROMATOGRAPHIC SEPARATION

Our sample for this injection will be a mixture of four phthalate ester standards dissolved in methanol. We will make 100× solutions of 500 μg/mL each of dimethyl, di-*n*-butyl, benzylbutyl, and di-*n*-octylphthalate. Next we will add a 1-mL sample of each compound to a 100-mL graduated cylinder and dilute with methanol to make a 5-ng/μL injection sample. The 100× sample should be centrifuged or filtered through a 0.54-μm filter. This same mixture will be used in Chapter 7 as a column quality control standard to study column performance.

We will inject 5 μL of our 1× sample into the injector using a splitless injection, injector temperature 275°C, and 35 cm/s He purge activation at 45 s. Both column oven program activation and mass spectrometer scan start are triggered by the injection of the sample. After the delay programmed to allow the injection solvent to pass through the mass spectrometer source, the filament will switch on and we will begin to see the TIC of our calibration test mixture on the data system. Figure 5.2 presents a chromatogram of a commercially available mixture that includes these four phthalates.

The first peak, methylphthalate, should come off after about 7 min and the fourth peak, di-*n*-octylphthalate, should elute after 10 min. The retention time and efficiency plate count of these two compounds should be stored for reference at a later time when we feel that the column's separating character has changed. We can pull the standards out of the freezer and rerun them as an independent check of column performance.

Phthalates-Column: DB™ -5.625

30 m x 0.32 mm I.D., 0.25 ym

Cat.No.:   123.5631
Carrier: Helium at 35 cm/sec
            measured at 50°C
Oven: 50°C for 1 min.
            50-200°C at 30°/min
            200-300°C at 15°/min.
            320°C for 4 min.
Injector: Spitless, 275°C; 45 sec.
            purge activation time
Detector: HD, 300°C
            Nitrogen makeup gas at
            30mL/min.

1. Dimethylphthalate
2. Diethylphthalate
3. Di-n-butylphthalate
4. Benzylbutylphthalate
5. Bis-[2-Ethylhexyl]phthalate
6. Di-n-octylphthalate

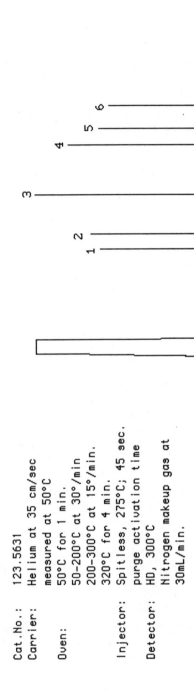

Figure 5.2   Calibration test mixture.

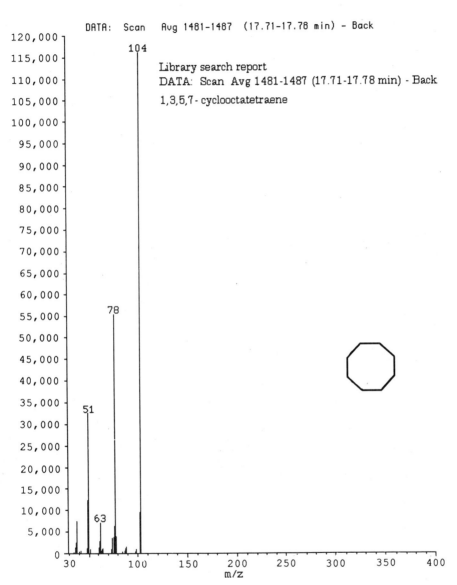

Figure 5.3   Spectral library mixture.

5.5. DATA COLLECTION PROCESSING

As we discussed in Chapter 1, the data system stores mass spectrometer data as a three-dimensional block with three axes: time, intensity, and mass–charge ratio (m/z); see Figure 1.6. The TIC in Figure 1.5 is a summation of the intensities of all mass fragments at a given time and is only one way of displaying the data in two dimensions. We can also make planar slices of the data and display them.

If we make our cut at a given m/z, we will display a single-ion chromatogram (SIC), intensity versus time; see Figure 1.7. At first glance the SIC appears to be similar to the TIC, but with fewer peaks. This is because the SIC is only displaying compounds containing the same mass fragment. If a compound does not contain that fragment, it will not appear in the SIC. The SIC is very useful for determining related compounds that break down to form common intermediates.

If we make a cut at a given time point, we can display this as a mass spectrum, intensity versus m/z; see Figure 1.8. The spectrum displays all the fragments associated with the chromatographic compound present at the selected time point. With these data we can determine the structure of the compound using mass spectral interpretation techniques. We can also use the data to identify a known compound by comparing it to a spectral library (Figure 5.3).

Library searching is done using probability matching usually starting with the largest peaks present and working down toward smaller peaks. The hit list contains all the possible matching spectra starting with the match of highest probability. Library software will usually display the spectrum, the matching spectrum that was found, a difference spectrum, and compound information for the matching compound, such as structure, molecular weight, other compound names, and physical data for the compound.

PART II

GC/MS OPTIMIZATION

CHAPTER 6

CHROMATOGRAPHIC METHODS DEVELOPMENT

Methods development in GC/MS has focused mainly on one variable, oven temperature control. A second variable used in GC methods development, column chemistry, has less impact on GC/MS due to the necessity of breaking and then restoring the MS vacuum in order to insert a new column. Column separation changes are primarily due to changes in the polarity of the packing materials in various columns. Nonpolar packings more strongly attract nonpolar compounds, which elute later than polar compounds. Hydrophilic, polar packings have more of an affinity for polar compounds, with nonpolar compounds eluting more rapidly.

Carrier gas flow rate and viscosity affect sample residence time on the column and column pressure. The faster the gas flows down the column, the shorter the sample residence time and the less chance compounds have to separate. Flow rate offers some potential for modifying a separation since it can be changed continuously. The limiting factor seems to be a rapid decrease in separation efficiency with increases and decreases in flow rate changes beyond an optimum flow rate.

Changing an injector throat liner can also affect a GC separation by increasing or decreasing the contact surface area, changing the character of the vaporized sample that actually enters the column. However, change cannot be made in a stepwise manner, and its effect is usually not predictable. Controlling the split ratio of a split–splitless injector can also affect the content of volatile material reaching the column head. But since it is controlled by the volatility of each component, the temperature of the liner, and the split residence time, it is very difficult to predict the exact composition of the injected sample. In actual practice both techniques are changed based on empirically derived information, and while their effects can sometimes be dramatic, they are at best unpredictable.

6.1. ISOTHERMAL OPERATION

In isothermal operation the column temperature is set one time and the column is allowed to reach a constant temperature before the sample is shot into the injector. Once a sample is vaporized in the injector and swept on to the column, it must interact with the column coating. This is often aided by running the column head at a lower temperature than the injector throat, partially condensing and concentrating the sample. The interaction of the sample at the top of the packing material also aids this concentration, leading to a sharpening of the disk of sample formed on the column head. An equilibrium is established between the concentration of each component in the coating on the packing and in the vapor phase above it.

This equilibrium is continuously disrupted in favor of the vapor phase component when the carrier gas sweeps down the column. As multiple interactions occur down the length of the capillary column, components with lower interaction affinity for the coating move more rapidly and begin to separate from the more highly retained material. Concentration disks of individual compounds begin to separate. As they travel down the column, diffusion effects, packing voids, and wall interactions begin to distort the shape of each separation disk. It broadens, the center tends to move faster than the outside edges, and it is pulled a bit into the shape of a nose cone. Finally, they reach the detector interface, enter the detector cell, and appear in the data system as a sharp front peak with maximum concentration at the peak center and some tailing on the backside as the trailing edges of the "nose cone" emerge. The longer a peak stays on the column, the broader its disk will become due to diffusion, but the more chance it will have to separate from other components of similar affinity for the packing.

Unless all components have a similar volatility at the column temperature, the later eluting components will begin to show broader and broader peaks, with more and more tailing, until the last components flatten down to the baseline. The separation still occurs, but not in a useful period of time. To speed this separation, we could start the column at a higher initial temperature. This, unfortunately, leads to a compression of the separation we achieved with the early eluting components. Fast-moving compounds do not have enough time on the column to fully separate. In the worst case, they will coelute as a broad, unresolved peak at the separation front.

6.2. LINEAR TEMPERATURE GRADIENTS

To resolve a mixture of compounds with widely differing retention on the packing, you need to run a linear temperature gradient. Equilibrate the column at an initial temperature that resolves early running peaks and then gradually increase the temperature to a final temperature that will remove all the components of interest. Hold at the final temperature long enough to resolve the last two peaks, then lower the oven temperature to the starting temperature for the next injection. It is important to hold the temperature at this point long enough to equilibrate the oven. Failure to do so will cause disturbances in the first part of the chromatogram; too low a temperature will cause early eluting compounds to retain longer. The more common problem is that the oven temperature will still be too high when the next injection is made and early runners will elute too early and be jammed together. If peaks are spread in the middle of the chromatogram, use a faster temperature rate increase to draw them together, although this may cause late running peaks to jam together.

6.3. ASSISTED REEQUILIBRATION

When you are trying to make as many injections as possible in the golden time between mass spectrometer tunings, eliminating dead times is critical to your success and profit margin. One of the major dead periods is the time required to cool and reequilibrate the GC oven.

The first method used to overcome this problem is to activate a piston used to push open the oven door at the end of the temperature ramp. The piston holds the door open until the starting temperature is reached and a spring pulls the door shut when the piston retracts. The next modification uses air blasting of the oven combined with the door opening to further accelerate the temperature drop. In the final method, cryogenic cooling from adiabatic expansion of compressed helium is added to further accelerate the cooling process on some machines.

Care must be taken that the cooling does not become too vigorous, resulting in extended heating time to reach the initial temperature. The reequilibration time is usually fast enough not to require modification.

6.4. HINGE POINT GRADIENT MODIFICATION

Next we must deal with compressed and expanded areas in the chromatogram. Compressed areas are sections of the separation in which poor separation causes many peaks to overlap. In an expanded area, peaks are overseparated and are increasing the total separation time. For each of these we can identify a "hinge point" before the first peak of the area (see Figure 6.1a).

Both areas can be handled by altering the rate of temperature change at these hinge points. The temperature change must be made while the sample is still in the column to have an effect on the elution pattern.

The first step in this hinge point development is to run a linear gradient and optimize the separation of early and late peaks using proper initial and final hold temperatures and times. Inspect the first compressed area in the chromatogram and determine the retention time of its first peak (t_{c1}). Measure the void time of the column by measuring the time from injection to the first peak front (t_0) for unretained solvent in the chromatogram (see Figure 6.1b).

Measure back from the start of the first peak in the compressed area an amount equal to the void time of the column ($t_{c1} - t_0$). Go to the column temperature profile program and enter a temperature hold starting at this time point with duration equal to the time width of the compressed area in the linear gradient. After this hold, enter a program step to return to the original temperature ramp rate. Reinject the sample and look at the effect this change produced (Figure 6.1c). Repeat the process for each compressed area in the separation.

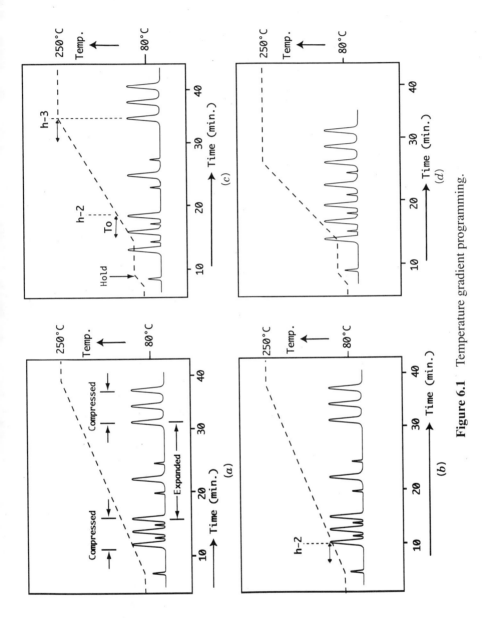

Figure 6.1 Temperature gradient programming.

For an expanded area of peaks, determine the hinge point, drop back the equivalent of the void time, and double the ramp rate over a time period equal to one-half the compressed area. Then, enter a programming step returning to the ramp rate and reinject (Figure 6.1d). Handle other expanded areas in the same fashion until the expected separation is achieved.

If you find that a temperature hold spreads the compressed separation too much, go back and replace the hold with a ramp with a slope halfway between the original ramp rate and the hold for the same duration of time. If doubling the ramp rate over a compressed area causes too much compression, pick a rate halfway between the old and new ramp rates and repeat the injection.

Remember oven temperatures turn like an 18-wheeler, not a sports car. They have to have time to produce an effect. Electrical Pelltier-type radiative heating and cooling of a much more confined column heater space might offer more precise control of both heating and reequilibration. It might also provide more rapid, precise temperature changes for developing fast, complex temperature gradients. At this time, I am unaware of any commercial GC systems using this type of column temperature control.

6.5. PRESSURE GRADIENT DEVELOPMENT

A new technique that shows promise for making controlled, predictable separation changes is electronic (column) pressure control (EPC). It was designed to correct for pressure drops occurring during injection that led to variations in sample run times due to variations in sample size.

Since pressure can be varied continuously and can produce changes in the time a sample stays on the column, it might be very useful as a control variable. Increase the pressure and increase the residence time. Decrease the pressure and decrease the time that samples have to interact with the packing materials and become separated.

Unlike column temperature control, which has lag time associated with heating and cooling the large volume of the oven, programmable pressure control should produce a nearly instantaneous effect on a separation. The questions about this technique are how wide a pressure range the capillary column will tolerate, the accuracy of the pressure control apparatus, and the linearity of the effect of pressure change on retention of compounds on the column. It certainly should be explored as a retention time variable for methods development and separation control.

6.6. COLUMN REPLACEMENT

Another area that could be explored for methods development is the rapid change of column types to produce an alpha effect on separation. Changes in column chemistry can make a dramatic change in separations, even change the order that peaks occur, aiding in building methods. Presently, to change a GC column, it is necessary to break the mass spectrometer's vacuum and shut the system down before inserting a different column. Because of the long delay in reestablishing the operating vacuum, column change is seldom used.

Rapid column changes would also be useful in troubleshooting system problems. Probably 70% of all system problems involve either injector or column contamination. Replacing the injector liner and the septum are easy and rapidly done. Ideally, replacing the column with a blank capillary column would allow a "column bridge" to quickly separate between column and hardware problems and do rapid system diagnostics.

To make this column change work for both methods development and troubleshooting, a method for rapid column replacement is needed. This could be done by building a simple column vacuum port similar to a DIP probe port. It might be possible to adapt some of the technology used in the septumless Nelson injector port to create this column port. A contaminated column should not require a 4-h shutoff for replacement when working in the 12-hour sample running time between EPA required tunings.

CHAPTER 7

MASS SPECTROMETER SETUP AND OPERATION

It would be nice if the mass spectrometer's mass axis came permanently calibrated. Instead, when the control system sends a specific DC/RF value to the analyzer, we have no way of knowing the m/z value of the selected ion fragment.

7.1. MASS SPECTROMETER CALIBRATION WITH CALIBRATION GASES

First we must analyze the fragments from a known calibration compound and then adjust the mass axis so it agrees with the expected fragment mass assignments. Periodically, we must go back and check to see that analyzer contaminations or degraded electronic components have not changed the selected mass positions. If the positions have moved, then we must recalibrate.

In order to obtain repeatable analysis from instrument to instrument or from laboratory to laboratory, we must also tune our instrument. We do this by adjusting various electrical components, such as the repeller, lens, and electron multiplier voltages, so that the mass spectrometer will give the expected relative ratios of ion fragment intensities for a target compound. Once target tuned in this fashion, two instruments should provide the same analysis for the same sample, no matter where the instruments are located.

Put another way, calibration adjusts the x axis of the spectrum so that we analyze the correct mass fragment at any point in a spectral scan. Tuning adjusts the analyzer so we see displayed on the y axis the same relative height, or abundance, of ion fragments each time we run the target compound. If it works for a selected known compound, it should work with a similar unknown within a specified concentration range. Each type of analysis has an expected target compound or compounds and an effective concentration range built into the description of the analysis.

Perfluorotributylamine (PFTBA or FC-43) is a clear, volatile liquid under the high-vacuum conditions of mass spectrometer analysis. It is kept in a vial valved off the sample inlet or the DIP probe port. When the instrument needs to be calibrated, the calibration gas valve is opened and calibration gas is allowed to vaporize into the source chamber. Calibration gas is ionized in the mass spectrometer's source by the electron beam from the filament and passed into the analyzer where its fragments are separated and detected.

The major ion fragment masses for calibration compound FC-43 are *69, 131, 219,* 264, *414,* 464, *502,* 614. In a well-tuned mass spectrometer, the 69 mass is the base mass; fragments 131 and 219 have approximately the same heights, equal to 45–60% of the 69 peak; the 414 peak is about 3–6% of the 69 peak; and 502 will be 3% or less of the 69 peak height (see Figure 5.1).

FC-43 has been the predominant calibration gas used in mass spectrometry because of the mass range of its fragments, their evenly spaced major fragments, and the volatility of the gas under the analyzer vacuum. Early quadrupole analyzers usually had a mass range of 0–800 amu. A few could reach 1000 amu and some research instruments offered an extended range of 0–2000 amu. These have become important in analyzing polymeric and multicharged molecules, such as peptides, proteins, and DNA fragments, with electrospray ionization. These extended mass ranges require calibration gases with larger mass fragments.

Perfluorotripentylamine and perfluorotrihexylamine offer fragments in the 500–600-amu and 700-amu ranges, respectively. Perfluorokerosene offers fragments well above 1000 amu. Perfluorophenanthrene is sold as a single, liquid calibration gas offering evenly spaced mass fragments from 50 to 650 amu. These calibration liquids are not as volatile as PFTBA, and may require heating of the holding vial for vaporization.

7.2. MASS AXIS TUNING

Modern high-performance control systems have autotune systems that both calibrate and target tune for specific target compounds. Their performance is so good that they almost never fail to reach the desired ratios. However, not everyone has updated equipment possessing this desirable feature, and some analyses require target tuning of compounds not listed in the control system. When in doubt, run one of the available target tunes first, then try your tune compound of choice to see if it also produces the correct fragment height ratios. If not, you are going to have to tweak the tune by manually adjusting the lens values or by using a lens scanner.

Autocalibration is the height of simplicity. To calibrate a mass spectrometer, set the mass spectrometer up for SCAN across the desired or specified mass range. Open the calibration gas valve and push the autotune button. The system will be busy for some time while it works its way toward the final autotune. Then it will display the resulting calibration spectra on the data display. You can usually select to print a calibration report that will include the spectrum and the numerical relative peak heights of each detected mass fragment.

Target tuning is autocalibration using calibration gas run to produce a specific target calibration. Using a different mathematical algorithm it produces the desired fragmentation ratios when the target compound passes from the gas chromatograph through the mass spectrometer's analyzer.

Once target tuned, the calibration gas is turned off. The mass spectrometer is set up in a SCAN mode, and a specified amount of the target compound is injected into the gas chromatograph. A tune report for the target compound should display the expected ratios of large and small peak masses. Usually this report is in the form of a pass–fail report for each fragment pair. If only one fragment ratio fails to pass, calibration parameters must be adjusted and the tuning compound reinjected until a pass is achieved on all specified peak ratios. Once this is achieved, the mass spectrometer is then certified for use for a specified period of time.

At the end of this time period, tuning compound must be reinjected and the certification report must pass again. If it fails, then the target tune calibration conditions must be adjusted, followed by injection of the tune compound until the system is recertified.

To manually calibrate a spectrum, set up a calibration scan from high to low mass and locate the 69 mass, which should be the largest mass present. Adjust it until its position is at 69. Check its separation from the smaller 70 mass; there should be little if any overlap. Now adjust the mass positions of the next two largest peaks, 131 and 219, and use the focusing lens to tune until the peaks are approximately equal in height, about one-half of the 69 peak height. If the peaks are much less than one-half of the 69 mass, increase the repeller setting until they are about equal to 50% of 69. Now you should be able to adjust the mass position of the 414 peak relative to the 219 peak. If you have problems finding the 414 peak, go back to the 264 peak and calibrate it against the 219 peak first. You usually will have no problem finding the 414 and 502 peaks after making that adjustment. Bring the electron multiplier (EM) voltage value up until the 414 peak height is about 4% of the 69 peak. You now should be able to see the 502 peak. Adjust its mass position. You can tweak its height with the amu offset (X-ray lens) or by increasing the EM voltage setting. For calibration, the 502 peak should be between 1 and 3% of the 69 peak. For use with most tuning compounds, it will be closer to 1%.

The 502 peak height is very susceptible to source contamination and is a good measure of when to clean the source. If the other peak heights are falling in the correct ratios but you cannot see the 502 peak, you are probably due for cleaning.

If you have to crank the EM voltage above 3500 without finding the 502 peak, you may also be in need of a detector replacement. I was told by a service man that the diagnostic test for the detector was to set the repeller to the maximum (10.0 V on a Hewlett-Packard mass spectrometer), open the calibration gas valve, and look at the 502 peak. On a new detector, the 502 peak should be 10% of the 69 peak height. If the value was below 5%, the detector should be replaced. He did not specify the EM voltage, but once a value of 3000 V is exceeded, the detector tends to degrade very rapidly. Keep the EM voltage only as high as necessary to see the 502 peak.

In a modern instrument, autocalibration is done first. Then, it may be tweaked manually by adjusting various lenses to achieve a specific purpose. For instance, you might use this technique in target tuning. Run a target calibration, inject the target compound, and find that several ratios are failing. Go back to the target spectrum and note that some fragments are too high. Turn on the calibration gas, adjust the focusing lenses, turn off the calibration gas, and reinject the target compound and see if the tuning report passes.

Older instruments using second-generation software often require considerable tuning expertise. Modern instrument systems are much easier to use, especially when the source is clean and the detector is new. In an operational environment, even they require a little tweaking. Experienced GC/MS operators earn their keep through the speed in which they can recertify a mass spectrometer's tune. Once the instrument is tuned, the operator is allowed to make as many runs as are possible before the next recertification. Run time equals increased analysis, which translates into dollars.

7.3. USING SYSTEM TUNING COMPOUNDS IN ENVIRONMENTAL ANALYSIS

The mainstay GC/MS analyses in the environmental laboratory are volatile organic analysis (VOA) and semivolatile organic analysis (semi-VOA). Both have tuning compounds defined that allow laboratories across the nation to report results that are reproducible. These analyses were developed by the U.S. Environmental Protection Agency for its contract laboratories program and picked up as standard procedures by other laboratories doing public contract analyses.

In using these compounds, the mass spectrometer is first calibrated to a specific set of tuning parameters and then the tuning compound is injected through the gas chromatograph. The spectra of the calibration compound is determined and height ratios of specified mass peaks are determined. All peak ratios must agree before the chromatographer can proceed with the analysis of standards and unknown samples. Failure of a single ratio to agree will require the analyst to return to calibration with PFTBA and retune the instrument, repeating injection of the tuning compound until agreement is achieved.

After all ratios pass, the instrument is certified for performance for a period of time specified by the method, usually 12 h. At the end of this time, the tuning compound must be reinjected and agreement of the tuning ratios again verified. Failing this, the instrument must be retuned with PFTBA until it again can pass a tuning report.

To run a bromofluorobenzene (BFB) tune for volatile organics analysis, set your GC oven to 230°C. Open the calibration gas valve and run a BFB target tune or ramp the entrance lens so that the 131 and 219 peaks are nearly equal, but slightly favoring the 131 peak. Ideal conditions are for the 69 peak, 100%; for 131, 35%; for 219, 30%; for 414, 1–2%; and for 502, 0.8%. Tweak the last two masses with your amu offset and the EM emission. Save your tune parameters.

Inject 50 ng of BFB solution. You should get a single peak at about 17.5 min. Figure 7.1 shows a spectrum and pass–fail tune report for BFB.

The EPA's Contract Laboratory Program (CLP) procedure requires that you select scans within 10% of the peak maximum. Other methods allow you to select any scans within the peak to pass BFB tune. Critical target mass ratios are 50/95, 75/95, 174/95, and the 174/175/176/177 quadruplet. The latter are particularly sensitive to high mass variations in the PFTBA tune.

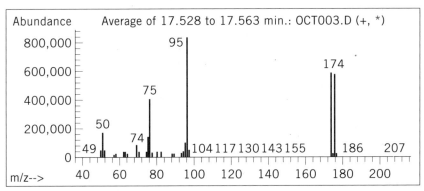

Peak apex is scan: 903

Target mass	Relation to mass	Lower limit, %	Upper limit, %	Relative abundance, %	Raw abundance, %	Result Pass/Fail
50	95	15	40	20.9	178,688	Pass
75	95	30	60	48.1	410,304	Pass
95	95	100	100	100.0	853,227	Pass
96	95	5	9	6.7	57,067	Pass
173	174	0	2	0.0	0	Pass
174	95	50	100	69.7	594,304	Pass
175	174	5	9	5.8	34,216	Pass
176	174	95	101	98.5	585,557	Pass
177	176	5	9	6.6	38,552	Pass

Figure 7.1   BFB tuning report.

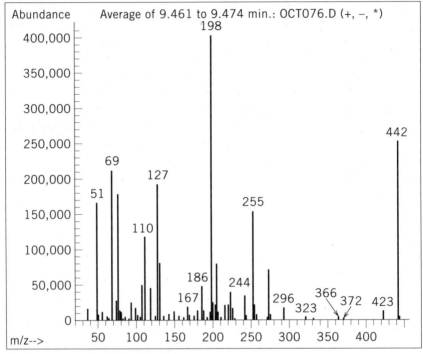

Peak apex is scan: 985
Average of 3 scans: 984, 985, 986 minus background scan 979

Target mass	Comparison mass	Lower limit, %	Upper limit, %	Relative abundance, %	Result Pass/Fail
51	198	30	60	41.7	Pass
68	69	0	2	0.0	Pass
69	198	0	100	52.4	Pass
70	69	0	2	0.0	Pass
127	198	40	60	47.9	Pass
197	198	0	1	0.0	Pass
198	198	100	100	100.0	Pass
199	198	5	9	7.0	Pass
275	198	10	30	18.2	Pass
365	198	1	100	1.9	Pass
441	443	0	100	0.0	Pass
442	198	40	100	63.2	Pass
443	442	17	23	19.7	Pass

Figure 7.2   DFTPP tuning report.

Semivolatile organics analysis uses decafluorotriphenylphosphine (DFTPP) as its tuning compound. The GC oven is heated at 250°C. The calibration gas valve is opened and DFTPP target tune is run on the calibration gas. If your mass spectrometer cannot do a target tune, ramp your entrance lens until the PFTBA 131 and 219 masses are approximately equal, then tweak the amu offset and EM voltage settings to reach a 502 value of about 1%. Ideal conditions for DFTPP tuning are 69, 100%; for 131, 35%; for 219, 30%; for 414, 1–2%; and for 502, 0.8%.

Once you have a satisfactory tune, inject 50 ng of the DFTPP solution. You should see a single peak at about 9.4 min. Figure 7.2 shows the spectrum and pass–fail report for a successful DFTPP tune.

The base peak for the analysis is 198. Critical target mass ratios are 51/198, 125/198, the 197/198/199 triple, 442/198, and the 442/443 doublet. The last two peaks are particularly sensitive to high mass variations in the PFTBA tune, source cleanliness, and detector aging.

7.4. ACQUIRING INFORMATION

The computer system for the GC/MS system has two functions. We have already discussed its use in controlling the programming of the gas chromatography, the autosampler, and the mass spectrometer. Its second function is to acquire data from the mass spectrometer, process it, and generate chromatograms, spectra, and data reports from the information.

The first decision you must make in setting up the mass spectrometer for data acquisition is selecting the operating modes. Most work is done in the EI (electron ionization) mode using SCAN.

The first decision must be the ionization mode. Changing from the EI to the CI (chemical ionization) mode involves dropping the mass spectrometer's temperature, breaking the system vacuum, disassembly the analyzer, replacing the EI source with a CI source, reassembly the analyzer, and reevacuating the system. This may take 4–5 h and is, therefore, not a trivial change to make. Newer systems and ion traps claim to be able to make this changeover without venting the vacuum. Many laboratories will dedicate a GC/MS system to CI mode molecular weight analysis to avoid having to go through this long conversion.

The change from the SCAN (mass axis scanning) to the SIM (single-ion monitoring) mode is not as physically challenging but is important because of the difference in the data produced. The SCAN mode is used when it is necessary to look at a large number of compounds in a single sample. It is the preferred mode when first examining a new sample or in doing methods development. You scan over the selected mass range at a scanning rate of about 30,000 points/s and then average the data obtained for each mass point. Although this appears to be a continuous scan, you are actually stepping from point to point. Step, settle, measure, step to the next point, and repeat. At the end of the scan range, you must make a big step back to the start of the scan range before starting the process again. For a mass range of 50–550 amu, measuring a point every 0.01 amu will obtain an average of 6 scans/s.

The SIM mode is chosen to look at a specific number of masses instead of every mass point within a given mass range. To look at four SIM peaks, you may only need look at 40 points/scan. Using the same scan rate of 30,000 points/s will average each point over 700 times in a second. You are measuring each point more accurately and this translates to an increase in sensitivity. The major use for the SIM mode, therefore, is to measure a limited number of masses at very high sensitivity. A second use would be to measure a limited number of peaks over a very long time (days or months) in order to decrease the number of stored data points. While monitoring changes in a gas stream, a customer scanned for 5 s once every hour for 5 months.

Once ionization and scanning modes are set, we are ready to begin acquiring data. The mass spectrometer is calibrated and tuned as described in the last chapter. Program written for autosampler and gas chromatograph control are downloaded or contact closures are sent out to begin remote programs in the individual modules. An injection signal is sent either from a manual injector or from the autosampler to start the dance. The gas chromatograph's oven temperature program begins to run and the mass spectrometer begins its scan program. A signal is sent to the A/D board telling it to begin acquiring analog data from the mass spectrometer's detector and convert it to digital data that is stored in a file on the computer's hard drive. For each acquisition time point, the entire averaged mass axis data scan points must be stored. If a data point is taken every millisecond, this represents a very large volume of data. A single 40-min GC/MS data set might occupy 2–3 MB of hard drive space.

7.5. DATA DISPLAYS AND LIBRARY SEARCHES

From this stored data, information is extracted for real-time displays. At each time point, the signal intensity of each mass point can be summed to give a total ion current. Display of the ion current at all the time points yield a total-ion chromatogram (TIC). You could also choose to display only the voltage signal supplied by a single mass ion for each time point as a single-ion chromatogram (SIC). These chromatograms can be displayed for either SCAN or SIM mode acquisition. The TIC for the SIM mode just contains ion current data summed from fewer mass points.

At a given time point, the intensities of the acquired mass spectrum points can be displayed as a mass spectra. While it would be possible to display each fractional mass point to generate a continuous spectrum, it is more common to sum all point intensities around a unit mass and display them as a spectral bar graph of discrete masses; see Figure 1.8. Since a spectra is distinctive and characteristic for each compound, it can be used as a fingerprint to identify it. The ability to be able to extract a spectrum on the fly allows us to identify compounds as they appear in the chromatogram.

Extracted spectra can then be submitted for library searching by comparison to a spectral database of known compounds. The results are displayed as a series of best matches with a confidence level assigned to each hit. Comparisons are usually begun with the major peaks in a spectra and move to lesser peaks. Failure to find an acceptable hit in the library databases are becoming increasingly rare as the databases grow. The 1996 NIST database of environmentally significant compounds contained 75,000 spectra (62,000 compounds), the Wiley Library (6th edition) contains 275,000 spectra, the Stan Pesticide Library lists 340 compounds, and the Pfleger Drug Library contains spectra for 4370 compounds. The NIST and Wiley libraries do not represent that many pure compounds since many are of the same compound run under differing mass spectrometer conditions. To be an exact fit, the mass spectrometers producing the target data must be calibrated and operated exactly the same as the sampling system. Multiple target spectra increase the probability of an acceptable hit. Chromatographic artifacts can also change the sample spectrum so that it fails to match the target. Trace artifacts can be removed by software that filters out minor sample mass fragments. Matching criteria can be modified by comparing the target spectra to the sample spectra instead of the other way around.

Failing to find a library match, the sample's fragmentation pattern can be examined and its structure determined by fragmentation analysis. This is a science unto itself and requires a chemist trained in the subject. A very brief introduction to structural interpretation is included in Chapter 11 and an excellent book on the subject by Fred McLafferty is referenced in Appendix D.

The GC/MS data set stored in the computer is a three-dimensional block of data. Each piece of information has three components: signal intensity, mass, and time. Some software offers topological displays of all the information in the array on a two-dimensional surface. Changes in signal intensity with mass and time are displayed as surface changes on the map's hills and valleys. While it is difficult to extract hard numbers from the map, it is very useful for observing trends, detecting impurities on peak shoulders, and predicting a compound's characteristic fragments in the presence of neighboring peaks. The latter will be very important when setting up target compound identification during quantitation.

CHAPTER 8

DATA PROCESSING
AND NETWORK INTERFACING

Once we have chromatographic peaks, the data and control computer can be used for compound quantitation. We can automatically determine the amounts of each compound present, positively identify known peaks, and refer unknown peaks to the library database for identification.

Some software allows control and data processing of multiple GC/MS systems on a single computer. A block diagram for a data/control computer system is illustrated in Figure 8.1.

The computer can also be connected to a computer network for data exchange and to avoid transcription errors from reentering information. When computer systems are retired and replaced, software exists to retrieve some of the archived data files stored in incompatible data formats.

8.1. PEAK IDENTIFICATION AND INTEGRATION

In automated target compound quantitation, you select and build a table of characteristic mass fragments and their relative signal strengths for each compound to be analyzed. The main identifier mass fragment for a compound is called the target ion; other identifier mass fragments for the same compound are called qualifiers. When these mass fragments appear in the spectra of a compound with the correct retention time in the

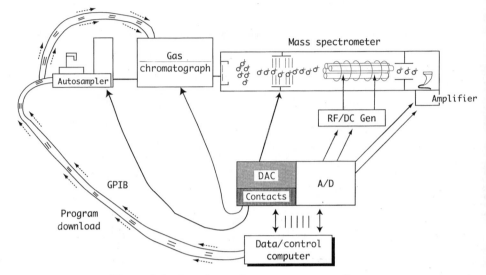

Figure 8.1 Data/control computer system diagram.

injected sample chromatogram, we have confirmed the identity of the targeted compound. You can gain further confidence when the relative mass intensity signals of the target ion and qualifiers agree with the expected target compound ratios.

Quantitation of the sample components can be done by comparing their target ion signal intensities to multilevel concentration curves for the target compounds. Internal standard compounds are added to the analysis mixture to correct for variations in injection volume and peak retention times on the column.

In environmental samples, we add surrogate compounds, physically similar to some of the target compounds, which are used to check for losses that may occur during extractions. Matrix blanks and matrix spikes are also included in our analysis deck to determine if materials are added or subtracted by the sampling matrix as a quality control check on the laboratory's technique.

In addition to target compounds, internal standards, and surrogates, we may find other compounds in the analyzed samples. Known compounds are compounds that we know to be in the original target sample but that we have chosen not to analyze. These are marked as such for sample analysis. Other compounds found in the final analysis sample that do not fit these categories can be referred to the library for searching as tentatively identified compounds for inclusion in a TIC report. Obviously, these cannot be quantitated, since calibration curves are not available for them.

In setting up a quantitation set, we first must calibrate the mass spectrometer and then run the tuning compound and adjust the lens until we have a tuning report. At that point, we begin to run a five-level quantitation set, each level containing target compound standards, internal standards, surrogates, and known compounds. If we have not built a table of target ions and qualifiers for each compound, we take a middle-level quantitation standard run, examine each target peak, surrogate, and internal standard, and select a target ion and qualifiers. When these are set, we run the matrix blank and matrix spike sample, and we are ready to begin to analyze real samples.

When we are through, the chromatograms and sample reports are examined. If compounds are incorrectly analyzed, because of retention time changes due to column contamination or temperature variations, they can sometimes be corrected by adjusting their position relative to the internal standard to which they are compared. In a mixture containing many compounds to be analyzed, multiple internal standards with differing retention times are included. Early running compounds are referenced to an early running internal standard, late runners to late running internal standards. By adjusting the internal standard's retention time, we adjust retention times for all associated peaks. Once all target compounds are analyzed, a series of reports can be generated.

8.2. MULTI-INSTRUMENT CONTROL

Modern computers are not limited to running a single GC/MS system. The minicomputer-based RTE system could control two GC/MS systems while acquiring data from both. Quantitation could be handled in a queuing system from data stored on the hard drive. Forms generation was a long, involved batch operation that ground processing and acquisition to a halt. Original specification for the RTE called for it to run four systems simultaneously, but this proved impossible in real-life situations. The latest generation of faster personal computers have made this possible.

To speed operations, many large laboratories divide this process across multiple computers, as computer prices dropped and speed and capacity increased. One computer was used to control multiple GC/MS systems. The data files, once acquired, were moved off to a second computer for quantitation that was not slowed by the necessity to time share system control and monitoring. A third computer could be used by a group assigned to monitor quality control and make chromatographic adjustment. Finally, a fourth computer could be assigned to do forms generation.

8.3. NETWORKING CONNECTION

At first, to make all this work, data was moved from computer to computer on diskette by "sneaker net." Once local-area networks (LANs) were set up, the data file could be moved electronically point to point from computer to computer. Eventually, with wide-area networks (WANs) in place, files such as reports and data sets could be moved from location to location, to computers of different types, even to facilities in other states.

Care had to be taken that data was not wiped from the original computer until it was certain that a copy had been received by the next processing computer. Archival storage of the original data set was important when data might have to be defended in court or before a government regulating organization. Large, sequentially stored tape data sets were built up to provide this legacy archive. These archives may never be needed, but when the data are required, they must be accessible. This has led to a potential problem facing contracting analytical laboratories, which will be discussed next.

8.4. REPLACEMENT CONTROL AND PROCESSING SYSTEMS

Over time, a problem arose with the GC/MS system. The mass spectrometer had changed little in the previous 15 years, except to become smaller and more compact. Compared to state-of-the-art computers, the heart of the processing and control system rapidly became outdated. Hard disks were too small to meet rising sample demand, the computers were too slow, and the programming was too difficult to use. In addition, the operating system and proprietary data storage system were incompatible with systems from other manufacturers. If a laboratory had bought systems from more than one vendor or different systems from the same vendor over time or acquired other laboratories with different systems, incompatibilities resulted. Only certain operators could run certain machines. If the operators were sick, the machines were sick. If the operator went on maternity leave, the machine was down or someone else had to be trained to run it.

Replacing computers with newer operating systems and better, faster software appeared to be the answer, but replacing a computer was like doing a heart transplant. The new software had to be able to control the mass spectrometer and, hopefully, the gas chromatograph and the autosampler. It had to be easy to use without requiring massive down time and retraining of the operator. Getting the computer was easy, but getting the software was another matter. Mass spectrometer manufacturers were only interested in manufacturing and selling complete systems. Until recently, few had any interest in resurrecting their older systems and upgrading the software that could be used on these systems.

Fortunately, a few third-party companies began to address this question. First, companies appeared that offered data processing on fast modern computers. You still had to run the old control system, but the data handling went much faster. In the last few years, full replacement systems have appeared that replace the control and data systems on almost any existing computer. They do full control of the gas chromatography and autosampler programming if these components are capable of remote control. And these system components do not all have to come from the same manufacturer, so you can upgrade other components, such as the gas chromatograph or the autosampler, as well as the computer. Some manufacturers have responded by offering upgrade data/control systems, but only for recent GC/MS systems and only for systems with modules that come from their company.

8.5. FILE CONVERSION AND DATA FILE EXCHANGE

Whether on a new GC/MS system or an upgraded system, with the new system came a new problem—actually an old problem recognized for the first time—data file incompatibility with the old data files in the archive. Even with a new system or an upgrade from the same manufacturer, data files would probably not be compatible. For example, the old files might be formatted in Pascal, Unix, or VMS, while the new files are stored in a Windows or DOS format.

How can the files be changed to the new format? Do you even want to convert them? What happens when the last tape drive and the hard drive that supports the old format bite the dust? Attempts to answer these questions have led to "elephant burial grounds" of old tape drives and obsolete computer control systems in many facilities trying to maintain access to valuable data archives. They certainly do not want to convert all the old, mostly obsolete data simply to have access to a single piece of data.

Again, third-party companies have come to the rescue. A software company named Chippewa Computing has produced software that converts Pascal-based formats to DOS for old Hewlett-Packard systems. A software package called Reflexions for the RTE, will convert its UNIX-based data format to DOS or Windows. There have been rumors of a similar package to convert Finnigan Mat DEC format to DOS, but this has not been verified.

In cooperation with the American Chemical Society, mass spectrometry organizations have been meeting to define a common database format called ansi-CDF. Each manufacturer would have the responsibility to write translators for each of their data formats into ansi-CDF. We could then translate data sets from one proprietary format to another by going through this "bridge format."

A similar CDF format proposed to all chromatography companies facing similar problems has been largely ignored. The key here is to follow the money. Companies have no financial incentive to cooperate and may open the door for a competitor into a laboratory they already control. The only hope of encouraging companies to provide common formats is for customers to make an issue of it.

Data reporting file exchange between data systems from different manufacturers also continue to be a problem in laboratories with GC/MS systems from different manufacturers and with systems of different ages from the same manufacturer. Environmental reporting requirements of the U.S. Environmental Protection Agency have forced a certain amount of standardizing in this field. They have set up reporting requirements for their contract laboratory program (CLP). A laboratory working within this program must submit these forms on paper and as disk deliverables. Equipment manufacturers and third-party software companies have written software designed to take the output from various GC/MS systems and processes it into CLP-type reports.

Because of the availability of this software, state and local regulatory agencies and commercial companies with internal environmental testing programs have adapted the federal requirements to their own need. Even though there were only 10 contract laboratories in 1995, currently these report types are being used throughout the United States and overseas. Often local modifications have loosened some of the requirements of the CLP for their own laboratories. The CLP reporting requirements have also undergone modification and fine tuning about every other year. But these remain the de facto standard for the industry and the same type of reporting can be produced from almost any GC/MS system.

8.6. DATA REENTRY AND TRANSCRIPTION ERRORS

We will assume that we can process the data into chromatograms and spectra and generate the required reports from the raw data after verifying the performance of the system. But, getting these data into a form for reporting to customers has proved a problem for commercial laboratories. The laboratory reports have to be combined with other analytical data on the sample and put in a form the customer can understand and use. The usual way that this is done is that the data from various analysis reports are abstracted and retyped into a final report format. The problem that often occurs is that data get reentered incorrectly. Transcription errors can be very expensive, in both time and money both to the customer and eventually to the commercial laboratory. Results have to be reexamined and corrected reports sent out. If this problem occurs too often, it is very easy to lose important customers in this very competitive industry.

To avoid, or at least minimize, the problem, many laboratories are moving to computer networked based laboratory information management systems (LIMS). LIMS is designed to prompt and check data type for manual input of sample information, such as source, ID numbers, and testing requirements. It then will pull information automatically from report files on a variety of laboratory computers to generate a final customer report without further operator intervention. Transcription errors are avoided, and once the system is up and running, report inspection is minimized. Setting up the system to extract exactly the needed report file components is much harder and time consuming than actually running the software.

A LIMS is only as good as its input data. It can detect missing information and flag it as an error, but incorrect data still generate incorrect final output. It is still necessary to have visual review and quality control of the data report information to assure the quality of the output going to the customer.

CHAPTER 9

SYSTEM MAINTENANCE
AND TROUBLESHOOTING

9.1. GAS CHROMATOGRAPH MAINTENANCE

The most important part of the gas chromatograph is the injection port. Approximately 90% of all chromatography problems can be traced to this component. This is where vaporization of the sample takes place. Maintaining inertness and a leak-free environment are top priorities in a laboratory where difficult sample matrices are analyzed (such as base neutral aromatics BNAs). The tasks described in this section should be performed daily.

9.1.1. Injector Port and Liner

Replace the injection port liner with a fresh liner daily. Liners should be silanized to remove active sites in the glass. Also a small "wisp" of silanized glass wool should be inserted into the liner about halfway. The glass wool increases the available surface area in the liner, which helps to promote vaporization. It also is effective in filtering nonvolatile residues and preventing them from making their way to the head of the column.

9.1.2. Septum Replacement

Every 2–3 days replace the septum. Septa are good for approximately 100 injections. If the septum is not changed regularly, it will begin to leak, causing retention time shifts and column bleed.

On Hewlett-Packard gas chromatographs replace the rubber O-ring around the liner and the injection port disk inside the nut at the base of the injection port.

9.1.3. Syringe Cleaning

Manual injection syringes should be rinsed at least twice with solvent before filling. Once a day rinse with an intermediate-polarity solvent [tetrahydrofuran (THF) or methylene chloride] to ensure that precipitated nonpolar solvents do not coat the barrel. If particulate from evaporation plug or block the syringe, ream it out with the fine wire supplied in the syringe box.

An autosampler syringe should be cleaned daily. Remove the syringe from its holder and rinse the plunger with methylene chloride. Also try to introduce some methylene chloride into the shaft of the syringe.

For column maintenance, I recommend using a guard column when possible. A guard column will collect nonvolatile residues that would otherwise accumulate at the head of the column. These residues can interfere with chromatography. If you do not use a guard column, trim 6–12 centimeters off the head of the column daily. Also, be sure to use a new ferrule when reinserting the column into the injection port. The column should be inserted 5 mm into the injection port.

Do not exceed the column manufacturer's maximum allowable temperature as this will cause column bleed and/or coating collapse, shortening the column's life. Also, guard against leaks, as oxygen will strip the stationary phase of the nonbonded column.

All fittings should be tightened one-half turn past finger tight. If you go much farther than this, the ferrule will fail and begin to leak.

9.1.4. Carrier Gas Selection and Purification

Helium is the most commonly used carrier gas, although it is expensive. It is inert and has low viscosity for good chromatographic separations. Hydrogen is cheap and has low viscosity, but it is also quite explosive.

Health and safety coordinators of most commercial laboratories will not approve its use. Nitrogen is inexpensive, inert, and nonflammable but has higher viscosity and thus is not an efficient carrier gas.

No matter what carrier gas you use, you will need to use an oxygen trap. The trap not only prevents trace amounts of oxygen present in the carrier gas from reaching the column, but also collects oxygen from leaks that may be present in the fittings. The trap should be changed at least every 6 months depending on the type of trap.

9.2. MASS SPECTROMETER MAINTENANCE

The two major problems in day-to-day mass spectrometer operation beyond reaching and holding a tune are air leaks and source burning. Most of your attention will be devoted to the instrument's source. Instrument manufacturers all have taken their own approach to source design, the objective being to create ions and to tune the source lens so that desired relative peak mass ratios and resolutions can be achieved. But no matter what instrument you have, you will have to clean the source from time to time. How often depends on the rate of use, the nature of the samples you are analyzing, and the frequency specified by your protocol.

9.2.1. Problem Diagnostics

If you are having trouble reaching full vacuum but the filament ignites, check for air leaks by scanning for m/z from 0 to 50 amu. Look for the water (18), nitrogen (28), and oxygen (32) peaks. If they are present, there is probably a leak around the column-to-source seal. Shut down and check the fitting and ferrule. If the fitting is snug, the ferrule is probably scored and needs replacing. Also check the seal around the calibration gas valve. Occasionally there are air leaks here. Often briefly opening the valve and then resealing it can eliminate this source.

An excellent method of determining when to do source maintenance is to monitor the 502 fragment height from autotuning. Measure the height after tuning a new or freshly cleaned instrument. Set an acceptable threshold, say 10% of the clean level. Once the 502 value drops below this minimum standard, it is time to clean. The 502 fragment is chosen because heavier fragments are much more easily effected by dirty or corroded source surfaces.

Once you have decided to clean the source, vent the analyzer, power down, and cool the gas chromatograph. Next, remove the column and the interface from the mass spectrometer. The analyzer source assembly is removed from the analyzer and disassembled. The ion and filament contract surfaces must be cleaned and dried. The ion source assembly is reassembled and inserted back into the analyzer. The interface and column are reconnected and the analyzer reevacuated and all temperature zones reheated. Finally, an autotune is run to establish that the 502 fragment height is back over the performance threshold.

The venting and power-down sequence will vary from instrument to instrument. It is important to follow exactly the procedure indicated in the instrument manual. Turbomolecular pumps are designed to be turned off at speed and vented through the rough pump. Oil diffusion pumps must be cooled to less than 100°C before venting or they will backstream oil into the analyzer, contaminating the quadrupole surfaces. Once cooled, carefully feel the temperature of the pump exhaust; it is a reliable guide to whether the diffusion pump is cool enough to turn off.

Once the GC oven is cooled and the column and interface removed, it is important to protect the interface insertion surface. Remove it from the analyzer and protect it by wrapping it in aluminum foil until it is time to insert it back into the analyzer section.

9.2.2. Source Cleaning

To clean the source, disassemble it in a clean area where there is plenty of room to work. Take special care not to lose small parts. Remove the control interface cables and the electrical connections to the filaments, the repeller, and the various focus lenses. Unfasten and remove retaining screws that hold the filaments and the repeller to the source body. On an HP 5972 mass spectrometer, the whole source assembly up to the entrance lens can be removed as a single piece for disassembly and cleaning. The pieces that need cleaning are those in contact with the ion stream: the repeller face, the ion source inner body, both sides of the draw-out plate and its pinhole entrance, the focus lens, and entrance lens contact surfaces. The ion source body shows burn next to where it contacts the filaments. This needs to be removed and the holes leading into the source body need to be cleaned by reaming with a fine drill.

Cleaning source surfaces is an art and a source of controversy. A variety of abrasive, chemical, sonic, and electroplating techniques have been described in the literature. Abrasive techniques using fine powders are effective, require minimum equipment, and are reasonably safe for the source surfaces. Hewlett-Packard recommends scrubbing these surfaces with an aluminum oxide powder and methanol paste for use with Q-tips. It also supplies aluminum oxide paper that can be used in cleaning inner surfaces and drill bits and a holder for reaming out pinhole entrances. Some laboratories consider aluminum oxide too harsh and use the less abrasive jeweler's rouge for cleaning. Also, there are laboratories that use a rouge paste in a tube that is intended for motorcycle detailing, and instrument service and repair facilities use high-pressure sand and water blasting techniques to clean the source.

Once the source elements are disassembled, the flat lenses can be cleaned lightly with a jeweler's paste and a Dremel tool. Under no circumstances should abrasive stones or rubber wheels be used. These have the potential of scarring the surface, which can alter the electrostatic field of the lens. Once this occurs, the element may not perform as well as originally intended. Parts that are too small for the Dremel tool should all be placed in a small beaker. The larger flat pieces should also be placed in a larger beaker. Fill the beakers with water and add a few drops of Alquinox (an industrial soap). Place the beakers in a sonicator and sonicate for approximately 1 h. After 1 h has elapsed, remove the beakers and very carefully pour out the water (being especially careful not to lose any parts). Then refill the beakers with methanol to remove the residual water. Sonicate again for about 5 min. Remove the parts from the beakers and dispose of the methanol in an appropriate waste solvent container.

Other rinsing procedures call for sonication in a series of solvents. After wiping all surfaces with Q-tips to remove as much abrasive as possible, it is recommended to sonicate for 5 min twice each in chlorinated solvents such as methylene chloride, then acetone, and finally methanol. Then air dry, place all parts in a beaker, and dry in an oven at 100°C for 15 min. Be very careful not to use abrasives on Vespal surfaces or allow solvent to get under these surfaces. They tear up easily and solvent will cause them to swell, making reassembly difficult. It is usually easy to recognize these colored, plastic-looking surfaces.

If you are in doubt, do not clean them. Burn and char are usually pretty obvious. Both types of char result from the high-temperature oxidation of nitrogen-containing organic compounds. The largest molecules in a separation exit the gas chromatograph at the hottest end of the temperature ramp. They then hit an evacuated volume where they are bombarded with 70-eV electrons.

Next, lay out the parts and inspect their cleanliness. You should have thin cotton gloves on to prevent finger grease from getting on the parts. If the source has small ceramic collars or spacers, inspect them for cracks or chips. It would be a good idea to replace these as necessary. Maintain a good spare-parts inventory to cover this. Reassembly is the reverse of disassembly. Reassemble the source body, including the insulators around the ion focus and entrance lenses. Insert the source body back into the analyzer body after reattaching the repeller and the filaments. Make sure that the ceramic collar between the source and the quadrupole does not bind; it must turn freely. Connect the repeller and filament leads as well as the leads to the focus lenses. When your source is assembled, you should conduct an electrical continuity check to ensure that, first, there is appropriate continuity and, second, no shorts exists. You are then ready to reinsert the analyzer into the mass spectrometer body.

Once the interface is reinserted and electrically connected to the gas chromatograph, reinsert the chromatography column, ensuring that proper insertion depth is achieved. How this is done depends on the source design. Generally, slide the column compression fitting and Vespal ferrule onto the column, insert it into the interface until it bottoms out, and retract the column about 2 cm. Tighten the compression fitting one half turn past finger tight.

Replace the housing on the vacuum containment vessel. Turn on the rough pump and begin evacuation. Set your interface heater to its operating temperature, usually around 280°C. Set the GC temperature zones to their startup values. When the gauge shows 10^{-4} torr, turn on the turbo or oil diffusion pump. If it is a diffusion pump, turn on the pump heater and bring it to the desired temperature. Pump until the normal high vacuum is reached, which will take about 4 h on an HP 5972 unit with a diffusion pump. This will vary with other systems.

If there is a problem establishing the rough vacuum, push down on the lid of the containment vessel. This is all that is necessary in most cases. If you still have a problem, stop the vacuum and inspect and clean the gasket around the inside of the lid. If necessary, replace the gasket.

Once full vacuum is reached and you have checked for air leaks by scanning below 50 amu, check the effectiveness of the cleaning procedure. Rerun the autotune procedure and check the height of the 502 peak. It should now be somewhere between the instrument's best measurement and the minimum system performance level.

9.2.3. Cleaning Quadrupole Rods

Another potential that can occur in the quadrupole that will affect its operation is accumulation of organics on the quadrupole rods. Ions that do not survive travel through the quadrupole collide with the rods, pick up an electron, and become electrically neutral molecules. If they are small volatile compounds, they are swept off the rod by the vacuum system and end their life as oil contamination in the vacuum pumps. However, there is a slow accumulation of larger, nonvolatile organics on the quadrupoles. Periodically, these must be washed off since they will distort the electromagnetic field and eventually shut down the analyzer. In order to remove the rods for cleaning, the system must be vented and the source removed as before. The ceramic collar between the source and the quadrupole is removed, electrical connection to the rods is removed, and the rod package is removed. The rods on most systems are held in exact hyperbolic alignment by two ceramic collars that *must not be* removed. Nothing will shut a mass spectrometer down faster than messing up rod alignments. The minimum that must be done is ship them in for repair and realignment. This is time consuming and expensive and not always successful.

Rods are cleaned by immersing the complete quadrupole unit of the four rods in their ceramic collars in a graduated cylinder and flooding it with solvent. You must be very careful not to chip a rod while placing it in the cylinder. Older quadrupoles have very large rod packages and are usually cleaned by wiping them with large lintless paper towels. Usually they are washed first with a nonpolar solvent such as hexane, then with methylene chloride, and finally with dry acetone. Modern quadrupoles can be air dried and then evacuated in a desiccator if the rod package is small enough to fit. Oven drying at 100°C has been rumored to cause rod distortion because of differential expansion of the collars and rods and probably should be avoided. Air dry large rod packages as much as possible; then finish the drying process as the rods are evacuated in the mass spectrometer. It will not help the roughing pump oil, but it will dry the rods.

9.2.4. Detector Replacement

Detector horns have a finite lifespan and should be replaced when noise begins to increase. Run the repeller to its maximum value, then look at the electron multiplier voltage necessary to get a 502 fragment for calibration gas above the benchmark value. When the EM voltage gets above 3500 V, it is time to consider replacing the detector.

9.2.5. Pump Maintenance and Oil Change

The roughing pumps are the only pumps you will be expected to service. Oil in these pumps should be changed every 6 months. Change the oil when it becomes brown and cloudy. You can observe this in the viewing port on the side.

Hoses coming from the roughing pumps should be periodically checked for cracks. Hose thickness and diameter are critical for proper performance of a vacuum system. Avoid the temptation to substitute other size tubing when you need to do emergency replacements.

Diffusion pumps should be serviced by trained technicians from the manufacturer. The same can be said about turbomechanical pumps. Most manufacturers offer some type of trading program for turbo pumps, and this should be part of the purchase agreement when buying a system. When these systems go down, rebuilding them is a major undertaking. They operate at very high speeds with little tolerance for error. When they are down, they are down.

9.3. SYSTEM ELECTRICAL GROUNDING

This is a problem that laboratories should never see. Grounding problems should be worked out by the manufacturers before and during system installation and should not occur unless the system is altered. However, systems do wear out, they are moved, and changes are made. When replacing the controlling software and interfaces and moving to modern computers, grounding problems can occur.

For a while I demonstrated replacement systems for a number of types of mass spectrometers. In a few cases, I saw problems with calibration gas peaks failing to stabilize. They jumped from side to side of the expected position. The problems only disappeared when the mass spectrometer chassis and the controlling interface were connected with a grounding strap and all electrical and computer systems were joined through a common surge protector. The problems seemed more common during winter months when laboratories are particularly dry. I suspect that static electricity discharges may be involved, since I have seen similar problems with other types of analytical systems during dry winter months.

A similar problem was seen and corrected on a system that was thought to have been subjected to a nearby lightning strike. The system had its original interface and computer system. The calibration problem was very similar to that seen on demonstration systems and was corrected by grounding the interface to the mass spectrometer. When this problem occurs it is very frustrating. You could almost calibrate the system, but it would never hold a tune sufficient for environmental analysis.

PART III

SPECIFIC APPLICATIONS OF MASS SPECTROMETRY

CHAPTER 10

GC/MS IN THE ENVIRONMENTAL LABORATORY

One of the major uses of GC/MS systems is in environmental testing laboratories. These can be commercial testing laboratories that do environmental testing for the public on a fee-for-test basis. They can be industrial in-house laboratories that do standard industrial testing as well as additional specific testing for their company's products and by-products. They can also be laboratories participating in the EPA's Contract Laboratory Program (CLP), which do testing for EPA regional laboratories. Most of these laboratories will be working with methods developed and standardized by the U.S. EPA for its CLP laboratories. Laboratories not involved with the CLP may modify the standard programs for their own use. Many will add compounds of interest to the lists of VOA and semi-VOA compounds. Some laboratories do not use the stringent tuning requirements of the EPA methodology. For example, the standard VOA analysis requires passing BFB tuning within the three scans closest to the BFB peak maximum. This is done to avoid contaminants that most often occur at the chromatographic peak front or on the tailing back side. Other laboratories only require passing a BFB tune with a scan anywhere within the BFB chromatographic peak. This probably seems a trivial change, but it can make a big difference in the time necessary to tune older mass spectrometers.

When you first look at environmental methods you are met with a bewildering list of numbered methods. Almost all are based on one of two procedures, with variations in the basic methods to allow for differences in the sample matrix. Volatile organic analysis is used to analyze low-molecular-weight compounds that can be purged out of aqueous solution with an inert gas and captured on a solid packing. Semi-VOA is used to quantitate larger, less volatile organic compounds that must be extracted from the matrix before they can be injected, separated in the gas chromatograph, and analyzed in the mass spectrometer. Volatile organic analyses are carried out on drinking water, wastewater, hazardous waste, and air-monitoring samples. Semi-VOA analyses are performed on drinking water, wastewater, and hazardous materials.

Other available GC/MS sample analysis methods used in some laboratories include dioxan/furan and pesticide/PCB confirmation. These represent a much smaller volume of work for the environmental laboratories. The analysis of dioxan/furan is also an extracted method and could be considered a semi-VOA analysis but is a very specific method with hazardous sample handling. It is not routinely done by all environmental laboratories. Pesticide/furan confirmation is an EPA-approved GC-only method; some laboratories used the GC/MS confirmation test to provide definitive proof that coeluting compounds are not confusing the GC-only method.

We will take a broad look at the VOA (method 624) and semi-VOA (method 625) of wastewater as representative of these techniques. For specific details of environmental analysis procedures, consult the latest U.S. EPA versions. These are available in print form directly from the EPA and in diskette and CD form from various suppliers for use with an analysis computer.

10.1. VOLATILE ORGANIC ANALYSIS: EPA METHOD 624

The EPA method 624 will be discussed as a typical VOA. It is an analysis of the volatile organic components of wastewater using a purge-and-trap apparatus to introduce sample into the GC/MS system. The compounds identified and quantified are halogenated hydrocarbons and aromatic hydrocarbons with boiling points below 200°C and molecular weights below 300 amu. Sample must be pulled, stored in amber glass bottles at 4°C, and analyzed within 14 days.

Method 524 for drinking water, method 8240 for hazardous waste, and the CLP 2/88 and 3/90 VOA methods are all similar with changes made for different sample matrix or reporting requirements.

In summary, the mass spectrometer running in EI mode must be calibrated along its mass axis using PFTBA calibration gas and tuned for BFB analysis using an autotune method or by hand if no autotune is available. The tune is checked by injection of a BFB solution through the gas chromatograph to see if the correct mass fragment ratios are produced (Table 10.1).

Once the GC/MS produces the BFB tune report, a standards calibration run must be made. If this is the first tune of the day, five concentration levels of the mixture of 624 VOA standards, surrogates, and internal standards are run and response factors calculated. If this is not the first tune of the day, a single-level continuing calibration (CC) standard run must be made and compared to the previous five-point standards run and should fall within acceptable variation ranges. If it fails to calibrate, then the five-point standards must be rerun.

Both surrogates and internal standards are spiked into the sample to be purged. Surrogates are compounds with chemical structures similar to the standard compounds, usually deuterated or fluorinated compounds, that are not found in nature and are added as a check on purging recovery. Internal standards are compounds with similar chromatographic behavior not affected by the method or matrix. Response factors for each compound are calculated for each target compound relative to the internal standards from the calibration standard runs.

TABLE 10.1   BFB *m/z* Abundance Criteria

Mass	*m/z* Abundance Criteria
50	15–40% of mass 95
76	30–60% of mass 95
95	Base peak, 100% relative abundance
96	5–9% of mass 95
173	<2% of mass 174
174	50% of mass 95
175	5–9% of mass 174
176	95% but <101% of mass 174
177	5–9% of mass 176

When acceptable standards are achieved, then blanks and samples can be run. A 5-mL analysis sample or blank is placed in the purge-and-trap apparatus illustrated in Figure 2.2 and sparged with 40 mL/min of helium for 11 min at ambient temperature into a baked sorbent trap predried with purge gas.

The trap is packed from the inlet end with equal amounts of Tenax (a porous, crosslinked resin based on 2,6-diphenylene oxide), then silica gel, and then charcoal. Organics are trapped in Tenax; water is trapped in the silica gel. The 7-1-87 EPA protocol adds a 1-cm plug of methyl silicone (OV-1) before the Tenax to protect and refresh the column but omits the charcoal plug at the end. Unretained purge gas is vented to the atmosphere through a vent valve.

After trapping is complete, the purge gas diverter valve is turned and helium is introduced from the exit end of the trap. The trap is rapidly heated to 180°C and back flushed with purge gas to desorb the organic components into the gas chromatographic column. The water trapped in the silica gel explodes into steam, helping to desorb the organics from the Tenax, but must be diverted away from the GC column with a valve into a dryer.

The GC column specified for the method is 1% SP1000 on Carbopack-B packed in a 6 f × 0.1 in. column. However, in recognition of advanced chromatography techniques, capillary and bonded-phase columns can be used. In selecting these methods, the analyst must adjust the analysis conditions to bring the method to compliance with expected separation standards. Due to the chemical stability of bonded-phase capillary columns, most analysts are moving to these columns to avoid column bleed into the mass spectrometer.

A GC oven temperature ramp program is run to elute chromatographic peaks into the mass spectrometer where they are scanned for identification and quantitation. The injector temperature is set at 200°C; helium carrier gas flow is adjusted to 10 mL/min. The initial GC oven temperature of 35°C is held for 6 min; then a 10°C/min ramp is run to 210°C, where it is held for 5 min. The sample passes through the interface heated to 200°C into the mass spectrometer EI source operated at 70 eV, which is scanned from 45 to 300 amu.

Table 10.2 shows VOA standards: retention times on a DB624 column 75 m long with 3.0 mm film thickness. Also shown are the primary ions, and experimentally determined minimum detection levels (MDLs) calculated from seven replicates.

TABLE 10.2 VOA Target Compound

Compound	Retention Time (min)	Primary Ion	Secondary Ion(s)	MDL (μg/L)
Chloromethane	7.84	50	52	2.1
Vinyl chloride	8.33	62	64	1.6
Chloroethane	9.99	64	66	1.3
1,1-Dichloroethene	12.59	61	96, 98	3.0
Methylene chloride	13.88	84	49, 86	2.1
1,2-*Trans*-dichloroethene	14.58	61	96, 98	1.6
1,1-Dichloroethane	15.53	63	65, 83	1.6
Chloroform	17.38	83	85	1.6
1,1,1-Trichloroethane	17.83	97	99	1.4
Carbon tetrachloride	18.18	117	119, 121	1.5
Benzene	18.52	78	77	1.6
1,2-Dichloroethane	18.53	62	64, 98	3.0
Trichloroethene	19.69	130	95, 132	2.1
1,2-Dichloropropane	20.11	63	65	1.8
Toluene	21.99	91	92	1.9
1,3-Dichloropropene	21.34	75	110, 112	2.1
1,1,2-Trichloroethane	22.60	97	83, 85	2.5
Tetrachloroethene	23.00	166	129, 164	1.9
Chlorobenzene	24.52	112	77, 114	1.6
Ethylbenzene	24.64	91	106	2.2

While the run is being made, the purge residue is flushed out of the purge tube with purge gas, rinsed two times with reagent water, and blown dry. The trap is baked out at 180°C with fore flow of purge gas to the vent, preparing it for the next sample.

The oven temperature must be returned rapidly to the injection temperature and equilibrated for the next injection. Automated purge-and-trap apparatuses are available from a number of companies so that a series of standards and/or samples can be prepared for sequential analysis.

The quantitation software database is prepared ahead of time with data from a middle-range calibration standard. Response factors and retention times are calculated from standard runs, and primary and secondary target ions (see Table 10.2) for each target compound are entered. Data from sample and blank runs are processed with this information to determine the identity and amounts of each target compound present. Known compounds, which will not be quantitated, surrogates, and internal standards are all marked in the quantitation software. Amounts of target compounds in matrix blanks, reagent blanks, QC check samples, and matrix spike samples are calculated for various quality control reports. Unknown compounds that are not target compounds, surrogates, internal standards, or known compounds are identified, referred to library searching, and reported as tentative identified compounds in a TIC report.

Basic total ion chromatograms need to be inspected by quality control before final reports are made. Compound retention times are required to be within a 30-s window, target masses must maximize within one scan of each other, and relative fragment mass peak heights must fall within 20% of those for a reference spectrum. When calibration standards are run, retention times can be adjusted relative to the retention of internal standards to correct for variations in the column separating characteristics. If sample or blank retention times fall outside these windows, the column must be modified or replaced and standards rerun.

Quality control is very important in these analyses. An initial method blank and standard spike in reagent water is required to demonstrate the laboratory's cleanliness and ability to run these analyses within parameters. Samples must be spiked with standards and reanalyzed for 5% of the total samples as a performance check. Check standards available from the EPA must be run periodically to demonstrate the laboratory's capability. A schematic of the method 624 VOA analysis is shown in Figure 10.1.

10.2. SEMI-VOLATILE ORGANIC ANALYSIS: EPA METHOD 625

Semi-VOA analysis is done by extracting base-neutral compounds from the grab sample with methylene chloride after pH adjustment to >11 with sodium hydroxide followed by acidification with sulfuric acid to pH < 2.0 and reextraction with methylene chloride. Grab samples must be stored in glass containers at 4°C and extracted within 7 days and completely analyzed within 40 days.

Method 625 is for wastewater samples, 525 for drinking water, and 8250/8270 for hazardous waste. The CLP methods for these were modified in February 1988 and March 1990 with another method change due soon.

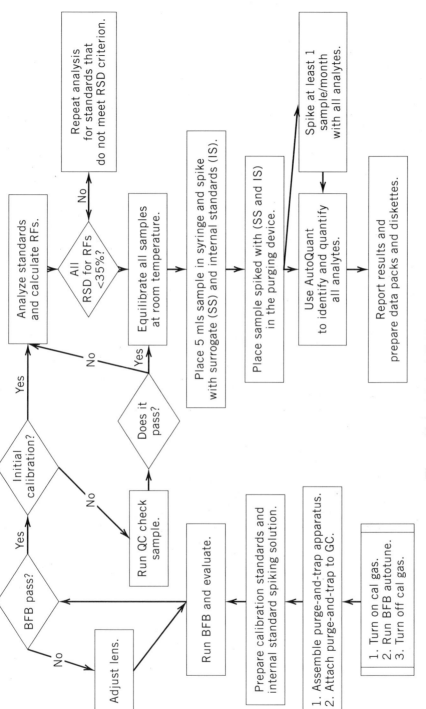

Figure 10.1 Method 624 VOA schematic.

Base-neutral compounds analyzed by method 625 are halogenated aromatics, nitro aromatics, polynuclear aromatics, aromatic ethers, pesticides, and PCBs. The possible presence of dioxins can be analyzed as part of this method, but they must be conclusively determined using EPA method 613. Compounds analyzed as acid extractables are substituted phenols. All compounds must have fragment masses below 450 amu.

Sample preparation is done by adding surrogate compounds to the analysis sample and blanks. The pH is adjusted to >11 with 10 N sodium hydroxide and the sample is extracted twice with methylene chloride to yield a base-neutral fraction. The aqueous fraction is then acidified to pH < 2 with 50% sulfuric acid and extracted twice more with methylene chloride to yield an acids fraction. The base-neutral and acids fractions are each dried separately over sodium sulfate and concentrated using Kuderna–Danish evaporators to remove solvent. Internal standards are added to the concentrates, which are made up to injection volume and placed in autosampler vials.

After the mass spectrometer is tuned using a decafluorophenylphosphine (DFTPP) autotune, it is set up for scanning from 35 to 450 amu and a sample of DFTPP is injected into the gas chromatograph. A DFTPP tune check is run using the abundances in Table 10.3.

If the tune check passes all criteria, base-neutral calibration standards or a continuing calibration standard is run. If the tune check fails, the lenses are readjusted and the DFTPP injection repeated until the DFTPP check passes. Once passed, the GC/MS system is certified to run samples for 12 h, after which it must be recertified with DFTPP.

The column used for base-neutral extractables is 1.8 m long by 2 mm inside diameter packed with 3% SP-2250 on Supelcort support. The method requires that a sample of benzidine must be injected and a tailing factor calculated. The column must be replaced if the tailing factor criterion cannot be achieved. One of the advantages of the capillary column is that removing a small portion on the inlet end of the column can aid in passing the tailing criterion and extend the lifetime of the column. If this is the first calibration, standards are run and response factors are calculated for all target compounds against internal standards.

The acidic extractable column is a 1.8 m × 2 mm column inside diameter packed with 1% SP-1240A on Supelcoport packing. It must be evaluated by injection with pentachlorophenol and checked for tailing. Once

TABLE 10.3   DFTPP *m/z* **Abundance Criteria**

Mass	*m/z* Abundance Criteria
51	30–60% of mass 198
68	<2% of mass 69
70	<2% of mass 69
127	40–60% of mass 198
197	<1% of mass 198
198	Base peak, 100% relative abundance
199	5–9% of mass 198
275	10–30% of mass 198
365	>1% of mass 198
441	Present, but less than mass 443
442	>40% of mass 198
443	17–23% of mass 442

it passes, standards are run and response factors calculated or a continuing calibration standard must be passed.

Now that standards are set, blanks and sample can be run on their respective columns. Base-neutral compounds are injected with helium carrier gas at 40 mL/min flow rate. The column equilibrated at 50°C is held for 4 min; then an 8°C/min ramp is run to a final temperature of 270°C and held at the final temperature for 30 min. Table 10.4 shows a list of the base-neutral extractables, their retention times, MDLs, and primary and secondary fragment ions.

The acid-extractable compounds are injected with helium carrier gas at 30 mL/min flow rate. The column is held isothermal at 50°C for 4 min; then an 8°C/min ramp is run to a final temperature of 200°C and held at the final temperature for 5 min. Table 10.5 shows a list of the acid extractables, their retention times, MDLs, and primary and secondary fragment ions.

As with the VOA sample, total ion chromatograms are examined to ensure that retention times, ion fragment masses, and peak heights are within acceptable limits. Any samples falling outside these limits must be reanalyzed. Quality control reagent blanks, spikes, and check compounds must be run periodically as in the VOA analysis. A schematic of method 625 semi-VOA analysis is shown in Figure 10.2.

TABLE 10.4 Semi-VOA (Base-Neutral) Target Compound

Compound	Retention Time (min)	Primary Ion	Secondary Ion(s)	MDL (μg/L)
1,3-Dichlorobenzene	7.4	146	148, 113	1.9
1,4-Dichlorobenzene	7.8	146	148, 113	4.4
Hexachlorothane	8.4	117	201, 199	1.6
Bis(2-chloroethyl) ether	8.4	93	63, 95	5.7
1,2-Dichlorobenzene	8.4	146	148, 113	1.9
Bis(2-chloroisopropyl) ether	9.3	45	77, 79	5.7
N-Nitroso di-n-propylamine	—	130	42, 101	—
Nitrobenzene	11.1	77	123, 65	1.9
Hexachlorobutadiene	11.4	225	223, 227	0.9
1,2,4-Trichlorobenzene	11.6	180	182, 145	1.9
Isophorone	11.9	82	95, 138	2.2
Naphthalene	12.1	128	129, 127	1.6
Bis(2-chloroethoxy) methane	12.2	93	95, 123	5.3
Hexachlorocyclopentadiene	13.9	237	235, 272	—
2-Chloronaphthalene	15.9	162	164, 127	1.9
Acenaphthylene	17.4	152	151, 153	3.5
Acenaphthene	17.8	154	153, 152	1.9
Dimethyl phthalate	18.3	163	194, 164	1.6
2,6-Dinitrotoluene	18.3	165	89, 121	1.9
Fluorene	19.5	166	165, 167	1.9
4-Chlorophenyl phenyl ether	19.5	204	206, 141	4.2
2,4-Dinitrotoluene	19.8	165	63, 182	5.7
Diethyl phthalate	20.1	149	177, 150	1.9

N-Nitrosodiphenylamine	20.5	169	168, 167	1.9
Hexachlorobenzene	21.0	284	142, 249	1.9
β-BHC	21.1	183	181, 109	—
4-Bromophenyl phenyl ether	21.2	248	250, 141	1.9
δ-BHC	22.4	183	181, 109	—
Phenanthrene	22.8	178	179, 176	5.4
Anthracene	22.8	178	179, 176	1.9
β-BHC	23.4	181	183, 109	4.2
Heptachlor	23.4	100	272, 274	1.9
δ-BHC	23.7	183	109, 181	3.1
Aldrin	24.0	66	263, 220	1.9
Dibutyl phthalate	24.7	149	150, 104	2.5
Heptachlor epoxide	25.6	353	355, 351	2.2
Endosulfan I	26.4	237	339, 341	—
Fluoranthene	26.5	202	101, 100	2.2
Dieldrin	27.2	79	263,279	2.5
4,4'-DDE	27.2	246	248,176	5.6
Pyrene	27.3	202	101, 100	1.9
Endrin	27.9	81	263, 82	—
Endosulfan II	28.6	237	339, 341	—
4,4'-DDD	28.6	235	237, 165	2.8
Benzidine	28.8	184	92, 185	44
4,4'-DDT	29.3	235	237, 165	4.7

TABLE 10.4 *(Continued)*

Compound	Retention Time (min)	Primary Ion	Secondary Ion(s)	MDL (μg/L)
Endosulfan sulfate	29.8	272	387, 422	5.6
Endrin aldehyde	—	67	345, 250	—
Butyl benzyl phthalate	29.9	149	91, 206	2.5
Bis(2-ethylhexyl) phthalate	30.6	149	167, 279	2.5
Chrysene	31.5	228	226, 229	2.5
Benzo(*a*)anthracene	31.5	228	229, 226	7.8
3,3'-Dichlorobenzidine	32.2	252	254, 126	16.5
Di-*n*-octyl phthalate	32.5	149	—	2.5
Benzo(*b*)fluoranthene	34.9	252	253, 125	4.8
Benzo(*k*)fluoranthene	34.9	252	253, 125	2.5
Benzo(*a*)pyrene	36.4	252	253, 125	2.5
Indeno(1,2,3-*cd*)pyrene	42.7	276	138, 277	3.7
Dibenzo (*a,h*)anthracene	43.2	278	139, 279	2.5
Benzo(*ghi*)perylene	45.1	276	138, 277	4.1
N-Nitrosodimethylamine	—	42	74, 44	—
Chlordane[a]	19–30	373	375, 377	—
Toxaphene[a]	25–34	159	231, 233	—
PCB 1016[a]	18–30	224	260, 294	—
PCB 1221[a]	15–30	190	224, 260	30
PCB 1232[a]	15–32	190	224, 260	—
PCB 1242[a]	15–32	224	260, 294	—
PCB 1248[a]	12–34	294	330, 262	—
PCB 1254[a]	22–34	294	330, 362	36
PCB 1260[a]	22–32	330	362, 394	—

[a]These compounds are mixtures of various isomers.

TABLE 10.5   Semi-VOA (Acid-Extractables) Target Compounds

Compound	Retention Time (min)	Primary Ion	Secondary Ion(s)	MDL (μg/L)
2-Chlorophenol	5.9	128	64, 130	3.3
2-Nitrophenol	6.5	139	65, 109	3.6
Phenol	8.0	94	65, 66	1.5
2,4-Dimethylphenol	9.4	122	107, 121	2.7
2,4-Dichlorophenol	9.8	162	164, 98	2.7
2,4,6-Trichlorophenol	11.8	196	198, 200	2.7
4-Chloro-3-methylphenol	13.2	142	107, 144	3.0
2,4-Dinitrophenol	15.9	184	63, 154	42
2-Methyl-4,6-dinitrophenol	16.2	198	182, 77	24
Pentachlorophenol	17.5	266	264, 268	3.6
4-Nitrophenol	20.3	65	139, 109	2.4

111

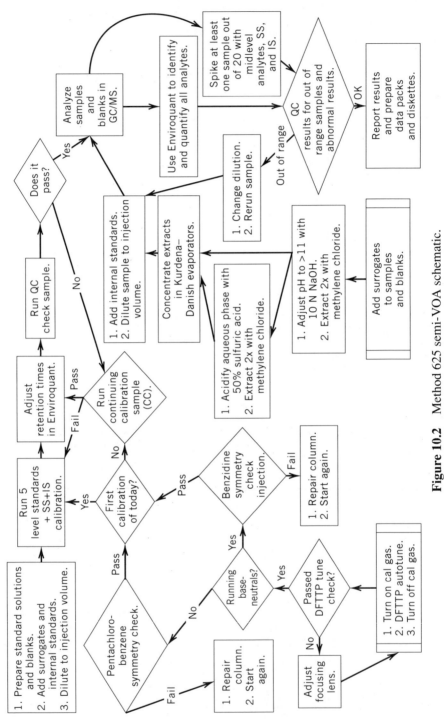

Figure 10.2 Method 625 semi-VOA schematic.

10.3. EPA AND STATE REPORTING REQUIREMENTS

When all samples have been properly analyzed and approved by quality control, the data are used to prepare the data package: BFB or DFTPP tuning reports, sample reports, surrogate reports, internal standard reports, blank reports, and TIC reports, if unknown compound analysis is required. The EPA requires that all CLP laboratories submit all of these reports along with the raw data in diskette form.

The EPA continues to modify its reporting requirements with modifications to the methods occurring every 3–5 years. Concurrent changes must be made to the quantitation software. Many CLP-like software packages are available, but they do not try to stay current with these changes. Various state and industrial laboratories modify the reporting requirements for their own needs. But many allow much more compact reporting and generally none are as strict as the EPA requirements.

CHAPTER 11

AN INTRODUCTION TO STRUCTURAL INTERPRETATION

I make no claims to being an expert in structural interpretation. This technique is a science in itself and beyond the scope of this book. The information presented in this chapter is merely the way I have used this technique in unraveling some questionable library assignments.

Interpretation of molecular structures from fragmentation data is an involved, time-consuming, and exacting science. If you are the type of person who enjoys doing the *New York Times* crossword puzzles, you might find it worth pursuing structural interpretation in more detail. McLafferty has an excellent book listed in Appendix D designed to help you learn to extract structural information using ion fragmentation mechanisms. Table 11.1 summarizes this method as well as the order in which you need to acquire information from the spectra.

When running a GC/MS system in the laboratory, how often is the rigorous interpretation of a structure needed? The answer is almost never! It is faster and a better use of your time to do a library database search and let the computer match the fragmentation pattern to know spectra. Does this mean that structural interpretation is useless? A dead science, of no value to the chromatographer? The province strictly of the mass spectroscopist? Not at all.

One of the problems with spectral library databases is that some of their structures are inaccurate. The original interpretation of their structures may

TABLE 11.1   A Guide to Molecular Structure by Fragment Analysis

1. Base Peaks and relative ion intensities:
   a. Determine molecular ion mass. Run CI if needed.
   b. A scarcity of major low even-mass ions = even-mass molecular weight (MW).
2. Elemental composition from isotopic abundances:
   a. Look for $A + 2$ pattern elements (Cl, Br, S, Si, O).
      (Check $A + 1$ ratios for absence/presence of S and Si.)
   b. Use the nitrogen rule to determine number of nitrogens.
      (If MW is even = 0 or even number of N. If MW is odd = odd number of N.)
   c. Number of C/Ni from $A + 1$ isotopic ratios.
   d. Estimate number of H, F, I, and P from A isotopic ratios and MW balance.
      (Only P is multivalent; F = 19 and I = 127 mass units.)
   e. Check allowance for rings and double bonds.
      [Number double bond or rings = $x - 1/2(y) + 1/2(z) + 1$,
      where $(C,Si)_x(H,F,Br,Cl)_y(N,P)_z(O,S)$.
3. Use molecular ion fragmentation mechanisms:
   a. Check fragment masses differences for expected losses
      (35 = Cl, 79 = Br, 15 = Me, 29 = Et, etc.).
   b. Look for expected substructures.
   c. Look for stable neutral loss (CH_2=CHR).
   d. Look for products of known rearrangements.
4. Postulate structures:
   a. Search library data base.
   b. Run hit compound on same instrument to confirm.
5. Use MS/MS if further confirmation is needed.

have been incorrect or mistakes may have been made in entering them. The previous Wiley database with 225,000 compounds was thought to have up to 8% incorrect structures in it. It is claimed that the current Wiley database has been cleaned up and that the structure assignments are >98% accurate.

Even when working with accurate known spectra and precise spectral matching algorithms there are still sources of problems:

1. The tuning conditions used in preparing the target spectra may not be the same as those being used in the laboratory.
2. The spectra could have been run on a different type of mass spectrometer with a different mass linearity. Some data in these libraries

were run on magnetic sector instruments rather than on a quadru-pole. The high mass areas of these two types of instruments do not calibrate the same way.

3. Either your spectra or the target spectra may have been run on im-pure compounds, which may introduce additional fragment peaks, especially at low relative intensities that may affect matching.

4. You may have chosen to scan above 40 amu to avoid water and air peaks, while the target spectra may include these extra fragments, altering the match.

The library matches do not give single compounds; rather they provide a list of matching compounds with a percent confidence level for that particular match. You may be provided with a pair or more of possible structures between which you will be required to choose. Partial structure interpretation can be a useful guide to making a choice between close matches or in determining whether a high probability match makes any sense at all.

11.1. HISTORY OF THE SAMPLE

The starting point for examining a fragmentation spectrum is to find out as much as possible about the compound being examined. Where did it come from? What kind of solubilities does it show? Does its UV spec-trum show conjugation or aromatic structures? What can the chromatog-raphy tell about its polarity? Does it show hydrogen bonding when run under conditions that break such bonds? With what kinds of compounds does it separate? What is its boiling point or melting point? What is its molecular weight? The more you know about the compound, the quicker you will be able to confirm the apparent match from the database.

Once we have its molecular weight and an idea of its chemical nature, we can determine its elemental composition from isotopic abundance in-formation calculated from fragment patterns. Finally, we examine the mass differences between fragments to determine what types of groups are being lost. If we do at least this much, we almost always have enough information to confirm a library structure assignment.

Molecular weight information is available from the compound's spec-tra. If the molecular ion is missing from the fragmentation pattern, being so unstable that it contributes at best only a very tiny peak, we can switch over and run in the chemical ionization mode. This will give us a molecular ion and, therefore, the compound's molecular weight as the first piece of information we must have to begin our analysis. The fragmentation pattern will tell us whether to expect an even or an odd mass molecular weight. If we look through our pattern and see a scarcity of major, even-mass ions in the low-mass range, we probably have an even-mass molecular weight.

11.2. ELEMENTAL COMPOSITION

Next we need to determine how many carbon, hydrogen, nitrogen, oxygen, and other elements are present. We can do this by looking for elements that show characteristic isotopic patterns in the fragment spectrum. Try to work with the most massive fragments and with the fragments having the tallest mass peaks. In any group, start with the most intense fragment group, the one with the most stable isotopes as the A peak.

Table 11.2 is a list of isotopic ratios for common elements making up organic molecules. A fragment containing an A-type element shows only a single band in the spectra. An A + 1 element, such as carbon or nitrogen, has two isotopic forms separated by 1 amu and forms pairs of fragment ions. The relative intensity peaks of the fragments will be the same

TABLE 11.2 Natural Isotopic Abundance of Common Elements

Element Type	Element	A		A + 1		A + 2	
		Mass	%	Mass	%	Mass	%
A	H	1	100	2	0.015		
	F	19	100				
	P	31	100				
	I	127	100				
A + 1	C	12	100	13	1.1		
	N	14	100	15	0.37		
A + 2	O	16	100	17	0.04	18	0.2
	Si	28	100	29	5.1	30	3.4
	S	32	100	33	0.79	34	4.4
	Cl	35	100	—	—	37	32.0
	Br	127	100	—	—	81	97.3

as the isotopic abundance of the element. If an ion fragment has a single carbon, the relative height of the first mass peak would be 100; 1 amu higher would be a fragment with a height of 1.1. The effect is additive. The more carbon atoms in the ion fragment, the higher will be the M + 1 peak height; that is, if five carbon atoms are present, the second peak would have a height equal to 5–6% relative to the first peak. If you do nothing else, find these A + 1 fragment pairs and use them to estimate the carbons present in the fragment. When you are working with organic molecules, you will be right much of the time. Biological molecules have enough nitrogen molecules to throw this number off.

Before we can work on carbon, nitrogen, and hydrogen, we must determine the presence or absence of other elements. Fragments containing the so-called A + 2 elements show a large peak, and depending on which element you are seeing, 2 amu higher shows a smaller peak of a precise height. Chlorine stands out like a sore thumb. A primary fragment peak containing chlorine shows a M + 2 secondary fragment one-third the height of the primary fragment. Every fragment containing a single chlorine will show the same 3:1 A + 2 ratio. This pattern occurs because chloride is a mixture of isotopes, its major isotope has mass 35, but it has a second major isotope with mass 37 with 32% of the 35 mass isotopic abundance. You can tell when a fragment decays with loss of this chlorine. The mass difference between fragments will be 35 and the A + 2 pattern will not appear in the smaller fragment. Bromine shows an A + 2 doublet of almost equal height (100 and 97%). Compounds with multiple chlorine molecules or a mix of chlorine and bromine in the same molecule show other characteristic patterns that are the additive results of combining A + 2 patterns. Tables of these are available in Throck Watson's book (see Appendix D).

Once we have determined the number of chlorines and bromines present and subtracted their contributions to the molecular weight, we need to look for the presence of sulfur and silicon. These are also A + 2 elements, but they show an additional isotope at the +1 position. First find the A + 2 patterns; then look at the mass in the middle. If there is no intermediate peak, scratch off these two elements. If there is an intermediate +1 peak, compare its height ratio to the A + 2 peak, after removing any chlorine and bromine, and examine the values in Table 11.2. This should lead to the number of sulfur or, less likely, the number of silicon present. Oxygen is also an A + 2 element, but the isotopic contribution from C^{18} is too low to be useful for measuring the amount of oxygen present in a fragment. Usually it is estimated from the residual molecular mass after the other elements (except hydrogen) are eliminated.

Once we have eliminated the $A + 1$ contributions from sulfur and silicon, we are ready to calculate the carbon and nitrogen values. In a simple hydrocarbon such as hexane, we should expect the $+1$ fragment peak to be about 6.6% as high as the main peak. Contributions by nitrogen are estimated using the nitrogen rule. This states that if the molecular weight is even, there either will be no nitrogen or an even number of nitrogens in the fragment. Odd molecular weights occur when there is an odd number of nitrogens. This allows us to either eliminate nitrogen or come up with a satisfactory number of nitrogens. Subtracting the nitrogen contribution should provide us with a good ratio of carbon isotopes, allowing us to calculate the number of carbons present. We can now subtract the carbon and nitrogen contributions to the molecular weight.

We are now left with hydrogen, fluorine, iodine, phosphorus, and, of course, oxygen. Because of its large isotopic mass, the presence or absence of iodine is usually obvious at this point and can generally be eliminated. Phosphorus is multivalent and most commonly bound to multiple oxygens and is usually easy to eliminate or identify. Fluorine's odd mass of 19 and its univalent replacement of hydrogen makes its presence or absence apparent when trying to distribute the residual molecular weight units between oxygen, hydrogen, and fluorine. Once an elemental assignment has been made (or even a partial that makes sense), check whether it agrees with the compound selected by the library search engine.

One more check that can be done is to check the number of double bonds and rings that are present. Table 11.1 presents a formula for calculating this number. Add up the number of quadrivalent, trivalent, divalent, and monovalent atoms present and plug them into the formula. The result represents the total number of double bonds and rings present using the lowest valence state for the elements. For a benzene ring this number is 4; for an electron balanced, charged ion this number might be $\frac{1}{2}$.

11.3. SEARCH FOR LOGICAL FRAGMENTATION INTERVALS

The final thing to look for in a spectra are mass losses between major fragment peaks. Look for characteristic losses like 35 (CL—), 15 (CH_3—), 29 (CH_3CH_2—), or a 15 loss followed by a series of 14 (—CH_2—), which indicates a breakdown of a straight-chair hydrocarbon. Also look for neutral molecule losses such as substituted vinyls ($RCH{=}CH_2$), which occur as part of rearrangements, and 28 (carbon monoxide), which may indicate the presents of a carboxylic acid or an aldehyde.

Once you find these markers, go back to the library structures and see if you can tell where these pieces are coming from. If none of these breakdowns make any sense, you may not have the right structure. If you can see how the pieces you are seeing can be formed, you have found confirmation for the structural assignment.

I hope this makes sense and helps you in confirming assigned structures. A rigorous study of fragmentation mechanisms will let your recognize more loss assignments, but you will have to determine whether it is worth your time.

In any case, the ultimate test is to acquire a sample of what you believe to be the correct compound. Run it on your GC/MS system with your tuning parameters under your chromatographic conditions to see if it gives the same spectrum.

CHAPTER 12

ION TRAP GC/MS SYSTEMS

Ion trap mass spectrometers are finding growing acceptance in GC/MS laboratories. Laboratories that use them claim they are 10–100 times more sensitive than a quadrupole. They can easily be switched between CI and EI modes, require less maintenance, and have potential to be used for MS/MS studies.

The desktop ion trap detector (ITD) and the floor-standing ion trap mass spectrometer (ITMS) vary in size and added function more than in theory of operation. The ITMS is designed as a research instrument with both analytical MS and MS/MS operation in mind. The ITD is a dedicated, compact unit with a smaller trap and pumping system designed for production GC/MS operation.

Molecules introduced into the ion trap are processed totally within the body of the ion trap. Uncharged material from the GC stream enters the trap around the ring electrode, is ionized, collides with other molecules, fragments, and is stored in stable orbits between the electrodes. The material is then eluted in increasing mass (m/z) by increasing the voltage on the ring electrode. This pushes each fragment ion into an unstable orbit, causing it to escape through one of the seven holes in the exit electrode and into the dynode electron multiplier detector, which sends a signal to the data system (Figure 12.1).

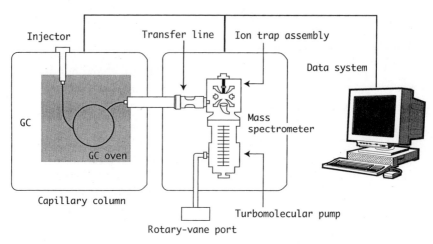

Figure 12.1   Ion trap GC/MS system.

12.1. ION TRAP COMPONENTS

The ITD GC/MS system is contained within two connected modules. The GC oven with the column, injector, and transfer interface is similar to the one used in the quadrupole system. The connection from the interface enters the ion trap through a transfer line just underneath the ring electrode. The detector horn lies immediately below the exit electrode.

The turbopump is mounted directly below and attached by a vitron O-ring gasket to the ion trap body, both of which are enclosed in a heated vacuum manifold. Also on the manifold are the attachments for the calibration gas valve, an entrance line for chemical ionization gas, and an exhaust port for the rotary vane mechanical pump.

12.2. ION TRAP OPERATION

Only a limited volume of sample can enter the ion trap without overloading and causing performance degradation. A narrow-bore capillary column with flows of 1 mL/min can be directly interfaced or a splitter column can be used to divert part of the GC stream to a secondary detector. Once the sample is in the trap, it is ionized with 70 V electrons from the ion gate in the entrance electrode at the top of the trap (Figure 12.2).

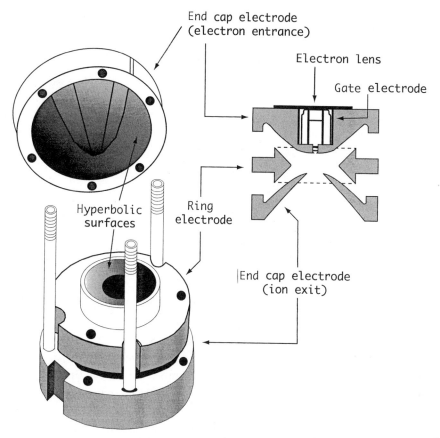

Figure 12.2   Ion trap electrode configuration.

Thermionic electrons are furnished by a heated filament. Between the filament and a unused spare filament is a repeller plate that drives the electrons toward the ion trap containment space (Figure 12.3).

At the base of the ionization electrode is a variably charged electron gate. When the gate has a high negative charge, electrons stay in the electrode; when the gate has a positive charge, electrons are forced into the storage space and ionize molecules of the sample.

The ring electrode around the containment space has a constant frequency, variable amplitude radio frequency signal applied to it. A storage voltage is applied to trap all ions with mass equal or greater than 20 amu.

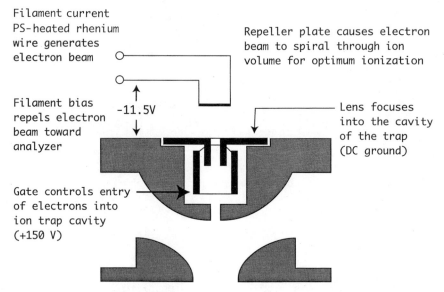

Filament current
PS-heated rhenium
wire generates
electron beam

Repeller plate causes electron
beam to spiral through ion
volume for optimum ionization

Filament bias
repels electron
beam toward
analyzer

-11.5V

Lens focuses
into the cavity
of the trap
(DC ground)

Gate controls entry
of electrons into
ion trap cavity
(+150 V)

Figure 12.3 Ion trap filament and ion gate.

At this voltage the ions formed are thrown into circular, hyperbolic orbits that are described as resembling the stitching on a baseball (Figure 12.4).

Approximately 50% of all ions formed are thought to be trapped and eventually reach the detector. This is compared to the single ion at a given time that reaches the detector in the quadrupole. Most ions end up colliding with the quadrupole rods and are never analyzed. This increased ion yield explains the increased sensitivity of the ion trap. Some increase in ion trap analyzer stability comes from the lack of sample accumulation on the electrodes, although this will vary from sample to sample. The helium carrier gas in the GC stream serves an important role in stabilizing the ions in their orbits. Frequent collisions between the small, fast-moving gas molecules and the charged ions dampens their movement, causing them to collapse toward the center of the trap.

The analysis is performed by gradually increasing the ring electrode *RF* voltage or scanning the voltage. This upsets the orbits of ions with increasing masses, causing them to escape through the exit electrode's holes and impact on the dynode's surface.

Figure 12.4 Ion trajectory schematic.

Ion orbital stability is also improved by applying axial modulation. This is a fixed frequency and amplitude voltage applied between the ionization electrode and the exit electrode at a frequency equal to about half that of the ring electrode. It has the effect of moving ions away from the center of the trap where the voltage is zero. This aids in ion ejection from the trap and dramatically sharpens the mass resolution at the detector.

Scanning is done in four segments over the full scanning range. This allows for mass peak-height manipulation and tune modification. With this tool, the tune can be adjusted to meet specific peak-ratios requirements.

The ion trap detector is the dynode electron multiplier previously used in quadrupole systems (Figure 12.5).

Positive ions striking the lead oxide glass cathode surface release electrons from the inner surface. These bounce down the inner walls, releasing a cascade of electrons on each contact; as many as 100,000 from a single positive contact will reach the anode cup and send a signal to the data system.

Ion traps are being used extensively in environmental analysis production laboratories. Their high sensitivity and easy maintenance make them attractive by avoiding downtime and providing trace analysis capability.

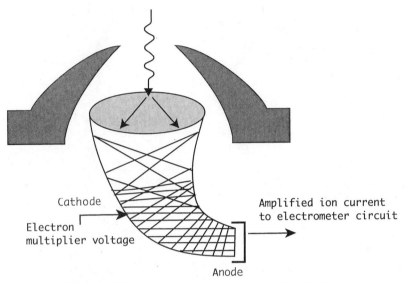

Figure 12.5 Ion trap exit electrode and dynode detector.

12.3. ION TRAPS IN THE ENVIRONMENTAL LABORATORY

Ion traps have received a bad rap in the past in that they do not produce spectra that match existing spectral libraries. This seems to have come about from poor autotune software in earlier traps. Using the four-segment scanning to balance tunes, they are able to meet environmental tuning specifications for BFB or DFTPP. They yield spectra that are easily searched and identified from either Wiley or NIST libraries. Libraries of ion traps only data are starting to emerge, with specific scanning parameters for ion traps, but these may prove useless wherever tuning compounds and parameters have been defined and reported.

12.4. CHEMICAL IONIZATION IN THE ION TRAP

One of the greatest but least appreciated advantage of the ion trap is its ability to do chemical ionization without switching the ionization source. Since ionization occurs in the trap itself, it is necessary to introduce only a primary ionizing gas such as methane, butane, or ammonia. Since the helium pressure is already about 10^{-3} and traps are susceptible to overloading, some loss of sensitivity may occur due to space charging of the trap volume. But switching is so easy that it is possible to program time of the jump from EI to CI in the same run. Since CI provides us with the molecular ion mass, it is a valuable aid in determining the structure for an unknown compound when combined with the fragment information from the EI run.

12.5. ION TRAP GC/MS/MS

Gas chromatography/tandem mass spectrometry is possible in an ion trap because alternate waveforms can be used to store specific selected ions in the trap. By allowing these ions to collide with themselves or with a heavy-makeup gas ion, such as xenon, the ion will fragment. The daughter ions produced can be used to help identify the parent ion and, by examination of fragmentations of a series of fragments, to identify related fragments from the original breakdown. In the next chapter MS/MS will be covered in more detail.

CHAPTER 13

OTHER GC/MS SYSTEMS

While quadrupole and ion trap systems represent the majority of commercial systems sold and used, other GC/MS systems are used for specific purposes. The first GC/MS systems were magnetic sector systems that generated an electromagnetic field to deflect ions into $m/3$-dependent curved flight paths. Due to their unusually stable magnetic fields, these systems have application in very accurate determination of molecular weights. Triple-quadrupole GC/MS systems are the favorite tool of research and methods development departments for determining structures of primary fragments. Time-of-flight (TOF) mass spectrometer systems have some general applications for volatile molecules but show growing use in LC/TOF-MS for molecular weight determination of large molecules such as proteins and DNA restriction fragments. The GC/FTMS system is a research tool offering the promise of provided much faster mass resolution and higher sensitivity than other existing techniques.

13.1. SEQUENTIAL MASS SPECTROMETRY (TRIPLE-QUADRUPOLE OR TANDEM GC/MS)

The so-called triple-quadrupole mass spectrometer is in reality an instrument made up of two scanning analyzers separated by a collision cell.

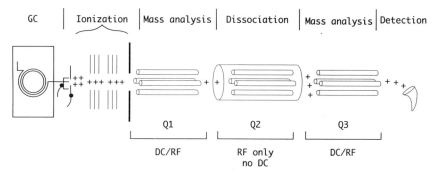

Figure 13.1   Triple-quad GC/MS system.

Fragments selected in the first quadrupole collide with inert gas, usually a large molecule like xenon, in the center quadrupole and undergo further fragmentation. The secondary fragments are then resolved in the final quadrupole analyzer (Figure 13.1).

The ionization source, focusing lens, and detector sections are identical to those in a single-quadrupole system. Collimator lenses after the collision cell focus the secondary ion fragments into the second analyzer. The purpose of the triple quadrupole is to allow separation and section of primary fragments in the first analyzer, fragmentation of the separated primaries in the collision cell, and analysis of secondary fragments in the second analyzer.

There are four possible modes of operation of the two analyzers: Q1, SCAN/Q3, SIM, called "daughter mode"; Q1, SIM/Q3, SCAN, called "parent mode"; Q1 SCAN/Q3, SCAN, referred to as "neutral loss scanning mode"; and Q1, SIM/Q3, SIM, referred to as "MRM, multiple reaction monitoring mode" (Figure 13.2).

The SCAN/SIM mode operation lets us determine which primary fragments are related to each other. The first quadrupole is scanned over a mass range and all fragments formed enter the collision cell and fragment to form secondary fragments. The third quadrupole is parked at a specific mass; only primary fragments that break down to form this specific secondary fragment will be detected. This common daughter ion points out interrelated primary fragments and helps us to understand which fragments are formed when a large primary fragment breaks down.

The SIM/SCAN operations parks the first quadrupole analyzer at a specific mass, allowing only a single primary fragment to enter the

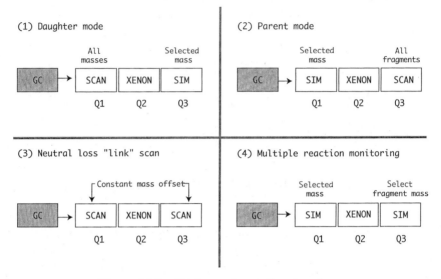

Figure 13.2   Triple-quad operational modes.

collision cell, where it fragments into secondaries. The final quadrupole is run in the full-scan mode, detecting all secondary fragments formed from this single primary parent, again providing structural information by showing its breakdown products.

The SCAN/SCAN operation is a little more complicated since both analyzer quadrupoles will be scanned at the same time but with a preset mass offset. When a primary fragment undergoes further fragmentation, it breaks into two pieces, a charged secondary fragment and a neutral molecule. What we are detecting in this mode are primaries that lose the same neutral molecule and therefore may be breaking down by the same fragmentation mode. The molecular mass of our suspected "neutral loss" is the value we assigned to our scan offset between the two quadrupoles. All primary fragments separated in the first analyzer enter the collision cell and fragment. Only secondary fragments whose mass differs from their primary fragment exactly the molecular mass of the neutral loss setting are detected by the final quadrupole. Any primary fragment that breaks down by forming a neutral molecule having a mass different from the offset mass will not be detected by the second analyzer.

The SIM/SIM operations is designed to definitely analyze specific components of very impure mixtures without having to completely purify them. Nature makes very complex mixtures that cannot always be completely separated either through extractions or by chromatography. We examine a chromatographic peak in which we expect a specific compound to appear by using the first quadrupole to separate a primary fragment characteristic of the compound of interest and pass it into the collision cell and use the final quadrupole to identify it by looking for only one of its specific daughter ions. We can identify and quantitate each targeted compound in a mixture, even if the chromatographic peaks that contain them are contaminated. For each compound to be analyzed, we select an individual primary and secondary fragment on a time basis in step with their expected chromatographic retention time.

Not all MS/MS systems use dual-quadrupole analyzers. As we mentioned in the last chapter, ion trap systems have built-in MS/MS capability. An ion trap/quadrupole system can be designed to pass specific parent ions to a quadrupole for daughter ion analysis. Or a specific daughter ion from ion trap MS/MS can be passed through the collision cell for quadrupole fragmentation analysis. Quadrupole/magnetic sector MS/MS combine the capabilities of both types of mass spectrometers: rapid scanning of the quadrupole for fragment separation and the stability of the magnetic sector secondary for accurate analysis of daughter ions. The literature contains examples of even more complex research systems: MS/MS/MS and even more exotic hybrid separating mode, multiple-analyzer systems, but these research tools are beyond the scope of this book.

13.2. MAGNETIC SECTOR SYSTEMS

The first commercially available GC/MS systems were magnetic sector instruments. They use an electromagnet based on a large permanent magnet to force ion fragments into circular sector flight patterns whose curvature is dependent on the fragment's m/z (Figure 13.3). The lower the m/z, the more deflection exhibited in the magnetic sector. Scanning of the mass range of ion fragments can be achieved in one of two ways: varying the accelerating voltage in the source or scanning the electromagnetic field strength of the magnetic sector. Detection is done by using a moving slit and a photomultiplier tube or by an electro-optical linear array.

The limitations of the magnetic sector systems are cost, size and weight of the permanent magnet, response time, sensitivity, and linearity, especially at the high-mass side. The spectra obtained are generally not

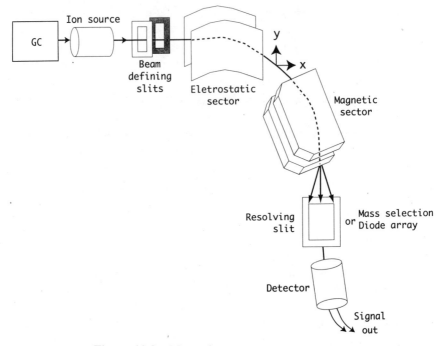

Figure 13.3   Magnetic sector mass spectrometer.

directly comparable to results from quadrupole or ion trap instruments for spectral extraction and library searching. Variation of the accelerating potential is limited in mass range and sensitivity drops off at the high-mass end. Scanning the magnetic field is the more commonly used technique, but it suffers from reluctance, an inertia resistance to magnetic field change. This leads to slower scan rates, which translates into poorer sensitivity. Much of this sensitivity problem is overcome in modern instruments by employing spatial array detectors so that the whole mass range can be measured at all times, which increases sampling rate and efficiency.

Magnetic sector instruments have made a comeback in the last few years because of their importance in accurate mass measurements for precise molecular weight determination. Injection is made from a probe rather than from a gas chromatograph. A technique called peak matching is used to compare the difference in accelerating voltage needed to make an unknown and a reference ion reach the detector at the same time. The reference ion must be within 10% of the unknown compound mass; masses can be measured to six decimal place accuracy with this technique.

Double-sector instruments are used to increase precision by using, in series, an electric sector to select ions of only one specific kinetic energy and a magnetic sector to peak match the reference and unknown compounds. Double sectors can be used for isotopic mass determination; they separate all ionic species into separate peaks that can be peak matched against a reference compound.

13.3. LASER TIME-OF-FLIGHT (GC/TOF-MS) GC/MS SYSTEMS

A growing segment of the GC/MS system market are using GC/TOF-MS systems. The TOF mass spectrometer uses a 3-kV electron beam to burst ionize in the mass spectrometer's source the sample from the gas chromatograph. The ionized fragments are repelled down a flight tube through a focusing lens.

The flight time of each fragment is dependent on its m/z, lighter fragments arriving first at the detector. To detect a given m/z fragment, the electron multiplier tube is activated only for a given time slice, allowing selection of only a single mass per burst. Flight time is very rapid, on the order of 90 ns for a 2-m flight tube. By increasing step-wise the time window for the electron multiplier for subsequent bursts, all masses over a given time range can be sampled and averaged fast enough to detect and analyze the narrow peaks produced by gas chromatography.

Since the majority of the fragments from every burst are discarded, sensitivity and resolution are potential problems. The SIM mode is the natural operating mode for a TOF instrument since the electron multiplier time window does not have to be changed. To increase sensitivity, a timed array detector is used in newer instruments. The array elements are set to sample the flight stream reaching the detector at different time windows. Using this technique, the whole burst fragment pattern can be analyzed for each event. Summing the resulting time windows allows a 10,000-fold increase in sensitivity. Arrays are limited by the number of array elements available to do the sampling and the inherent noisiness of the array. A 50×50 array provides 2500 sample points. For a 0–800-amu detection range, this provides a 0.3-amu resolution. Typical quadrupole resolution is 0.1 amu or better.

The length of the flight tube has historically produced very large, cumbersome TOF instruments. This problem has been reduced by folding the tube using electrical "mirrors" to reflect and accelerate the fragment

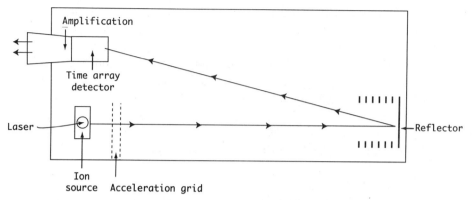

Figure 13.4   Reflectron time-of-flight GC/MS system.

flight stream back down the flight tube so as to impact the detector (Figure 13.4).

Time-of-flight GC/MS systems are rare outside academia. This MS technique is having some success in LC/MS, where LC/MALDI-TOF/MS systems are used for analysis of proteins, peptides, and polynucleotides. (Here, MALDI is short for maser-assisted laser desorption and ionization.) The liquid stream from the HPLC is mixed with a chromaphore, such as cyanocrotonic acid, which will absorb light from the high-intensity laser burst in the source. These target dye molecules explode, throwing the accompanying protein into the gaseous phase and at the same time chemically ionizing it. Since the free amino groups on the side chains provide multiple ionization sites, a series of multiply charged molecular ions each with a different charge are formed from a single protein. These are repelled down from the flight tube and separated at their m/z. Analysis of this family of molecular ions, which differ by the size of their charge z, allows calculation of the molecular weight of the original protein. Charges as large as 20–50 on an ion radical allow enzymatic size proteins (MW 25–60 kilodaltons) to be separated on a TOF mass spectrometer with a separating range of 0–1000 amu.

13.4. FOURIER TRANSFORM (GC/FT-MS) GC/MS SYSTEMS

Gas chromatography/Fourier transform mass spectroscopy (GC/FT-MS) produces mass spectra using ion cyclotron resonance (ICR). The sample is ionized by a burst of electrons in the source and passed into the analysis cell where they are held in a constant magnetic field provided by trapping plates. Each fragment will follow a circular orbit with a cyclotronic frequency characteristic of its *m/z* value.

To detect the fragments present, a full-frequency *RF* "chirp" signal is applied from a transmitter plate perpendicular to the trapping plate. Ions absorb energy from the chirp at their cyclotronic frequency and are promoted to a higher orbit. Detector plates perpendicular to the third plane

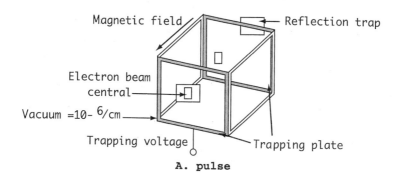

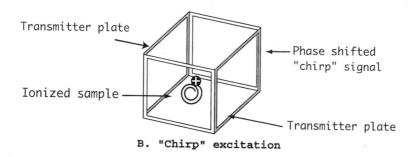

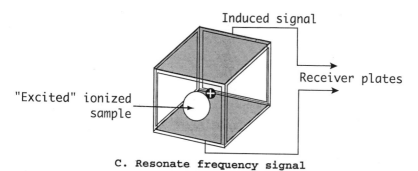

Figure 13.5   Fourier transform GC/MS system.

of the cell measure a complex signal containing all the frequencies of the excited fragments (Figure 13.5). Fourier transform software converts this frequency snapshot to a spectrum of the m/z values present in the sample. Like the spatial array detectors, every fragment is analyzed for every ionization burst event. Ionization can be equivalent to EI or CI, and examples of laser-assisted ionization exist in the literature.

Measurements can be made in milliseconds and have been used to monitor very short gas phase reactions. Since the ion fragments are not destroyed in the cell, multiple measurements over time can be averaged to produce a very accurate, high-resolution measurement yielding excellent sensitivity. The signal tends to be very stable and is not dependent on ion optics or variation in detector electronics. Modern computers can provide transformation calculation fast enough to provide real-time data.

CHAPTER 14

AN INTRODUCTION TO LC/MS

Interfacing a mass spectrometer to a high-performance liquid chromatograph, a most important potential addition to the HPLC arsenal, is not a
new technique. The mass spectrometer began as an outgrowth of the
Manhattan Project during World War II. As investigators involved in this
program returned to their respective universities, the techniques they had
developed, and in many cases the equipment, returned with them.

In the 1960s a GC/MS interface was developed, but the first HPLC/MS
interface did not appear until the 1970s due to the problem of detecting
compounds in the presence of large amounts of solvent. The mass spectrometer is very nearly the perfect HPLC detector since it allows noncon-
troversial identification of even "unknown" compounds from their fragmentation spectra. The problems preventing widespread introduction of
LC/MS into general laboratory use have been threefold: the price, getting
rid of large amounts of HPLC solvent, and expertise in interpreting results. Ten years ago the mass spectrometer system was a massive instrument costing in excess of $100,000 with an LC interface costing in ex-
cess of $20,000. The high-vacuum pumps required to run the *MS* system
needed constant maintenance. The high level of organic solvents in the
HPLC mobile phase tended to overload the mass spectrometer and over-
work the pumping systems. Interpretation of spectra was tricky and required some specialization in the field.

Many of these problems are rapidly disappearing. Desktop mass spectrometry detectors (MSDs) have shrunk in size and prices have dropped to around $48,000. Pumping systems are becoming less demanding of service. The solvent problem is finally coming under control using nebulizers and splitters. Interpretation of results is much easier and faster due to computerized on-line, rapid spectral library database searching.

While prices of new systems are still prohibitive for the average chromatography laboratory, older systems have been retrofitted with modern data systems, equipped with home-built LC interfaces, and put back into operation for around $30,000. As prices continue to drop and technology advances, LC/MS will become a major tool for the forensic chemist (analyzing drugs of abuse and diluents) and the arson investigator (analyzing the presence of accelerants at a fire scene), whose data must stand up in court. It will also be important to the clinical and pharmaceutical chemist, whose separations ionically charged, water-soluble compounds impact life and death. Food and environmental chemists must analyze samples that have an effect on the food we eat, the water we drink, and the air we breathe. All can analyze a broader range of material more rapidly without the limitations imposed by gas chromatography. Almost any compound that will dissolve can be separated in a high-performance liquid chromatograph and, potentially, analyzed by a mass spectrometer.

Let us now look at the design of the LC/MS system, its operation, and the way mass spectral data are manipulated to produce chromatographic information and compound identification. This will be simply an overview, as LC/MS is a field in itself. But it is important for the laboratory investigator to have a working knowledge of this technique and the future it promises.

14.1. LIQUID INTERFACING INTO THE MASS SPECTROMETER

The problem faced in the LC interface is introducing larger volumes of mobile phase along with the compound to be analyzed into the high vacuum environment of the mass spectrometer source. Liquid chromatography/mass spectrometry began in 1969 with a 1-μL/min flow into an EI source. The sample concentration was so much lower than the amount of solvent present that it was nearly impossible to detect the target masses of sample. Much effort was taken to increase sample concentration. Microflow systems and capillary columns were investigated as ways of increasing column efficiency, sharpening sample bands, and increasing concentration. These techniques were only marginally successful.

In the 1970s, an LC source was developed using a continuous, moving metal band that pulled a portion of the column effluent into a heated vacuum oven and then into the mass spectrometer source for ionization. Flow rates were reduced by using a splitter in the effluent line so that the capacity of the band was not exceeded. The system worked, but volatile components were swept away with the solvent and thermally unstable compounds degraded in the drying oven.

14.2. THERMOSPRAY LC/MS

The first modern interface for thermospray (TSI) LC/MS was introduced in 1983 and allowed introduction of column effluent at 1.0–1.5 mL/min (Figure 14.1). The mobile phase had to be mostly water because it had to contain large amounts of volatile buffer, in excess of 100 mM, to induce chemical ionization.

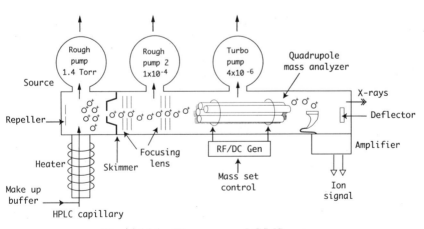

Figure 14.1   Thermospray LC/MS system.

The mobile phase was forced through an electrically heated capillary and out through a fine orifice into a high vacuum. Explosive last stage evaporation of the solvent drop in the presence of high concentrations of ionized buffer chemically ionized the sample. The ionized sample molecular ion was pulled by a voltage differential into the analyzer. Because of the low-energy chemical ionization mode used, only a small amount of sample was converted to ion fragment and the data produced were primarily used to provide information for molecular weight determinations.

That the entrance capillary must often be heated to 200°C and the requirement for 100 mM volatile buffer often lead to sample decomposition and orifice plugging. Roughing vacuum pumps and jet separators were added to a preheated interface to remove the large amounts of evaporated solvent. The original TSI system was designed for isocratic operation, but an interface was developed that allowed solvent gradient operation by postcolumn addition of compensating amounts of high concentration (100 mm) buffer solution. This makeup buffer had to be added to maintain the ion concentration of the solution. Above a certain level of organic solvent buffer precipitation could not be avoided.

14.3. ELECTROSPRAY LC/MS

The electrospray LC/MS system (ESI) shows tremendous application for producing multiply charged molecules. It has been applied to molecular weight determinations of protein and large peptides and shows promise for analyzing DNA restriction fragments. This system is limited to microflow applications (1–5 μL/min) in which effluent is forced through a capillary out into the source vacuum through a coronal electron discharge operating at 25,000 V. The discharge is produced off the sharp tip of a very fine needle. Electrons released at the needle tip form a cloud through which the mobile phase stream from a capillary tube explodes into the evacuated interface. These electrons knock other electrons off the sample, producing molecular ions. Mobile phase buffer can cause serious plugging of the capillary tip and corrosion of the coronal discharge needle tip and should be avoided in electron spray applications (Figure 14.2).

Proteins can acquire multiple positive charges at basic amino acids such as lysine, arginine, and histidine. Since the MS analyzer separates on the basis of m/z, or mass divided by charge, mass spectrometers with

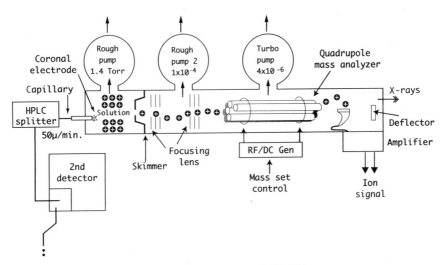

Figure 14.2   Electrospray LC/MS system.

an operating range of 0–2000 amu can still detect proteins with 10–50 charges per molecule.

Deconvolution of the charge envelope of multiply charged fragments developed by a single protein allows calculation of the protein's molecular weight. This can be estimated from knowledge of the fragment's charges and the mass difference between adjacent fragments with incremental charges. Software is available to detect all related charge pairs throughout the envelope and calculate average molecular weights. Using this software, it is possible to detect and determine molecular weights for coeluting proteins showing overlapping fragment envelopes.

14.4. ATMOSPHERIC PRESSURE INTERFACE LC/MS

The atmospheric pressure interfaces (APIs) for LC/MS show real promise for general HPLC application. Originally developed for protein electrospray applications, they employ a nebulizer sleeve around the effluent inlet capillary. Injection of an inert gas such as nitrogen into the nebulizer provides a high-velocity gas jet that breaks down the atmospheric effluent into a fine mist to aid in evaporation. In the heated nebulizer interface

(HNI) makeup nitrogen gas sweeps the tiny sample droplets into an electrically heated tube and then out over a coronal discharge needle. Charged sample ions are pulled by a voltage potential difference through an inert curtain gas into the evacuated source. From there the repeller forces them through the focusing lens and onto the analyzer rods.

14.5. ION SPRAY LC/MS

The ion spray interface (ISI) reverses the process. The nebulizer gas converts the effluent into a fine mist in the presence of a high electrical potential coronal discharge. These charged small drops are swept toward a grounded liquid shield on which large droplets impinge and run off. The fine, charged mist is pulled into an electrically heated capillary in a first-stage vacuum chamber leading to the mass spectrometer source. After the solvent evaporates, the charged molecules are pushed by the repeller and focused by electric lens into the analyzer and onto the detector (Figure 14.3).

Both interfaces allow high-sensitivity HPLC operation at 1–1.5 mL/min flow rate without a stream splitter, run gradient effluents without makeup buffer, and produce a CI molecular ion. Ion spray has the advantage of being able to produce EI fragmentation by increasing the voltage potential between the nebulizer tip and the liquid shield, effectively creating a poor man's MS/MS system. At low voltage mainly molecular ions are produced, yielding molecular weight information. At higher voltages frag-

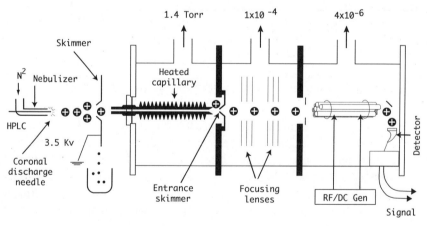

Figure 14.3 Ion spray LC/MS system.

menting occurs to produce an EI-type pattern. A modification of this system switches voltage from high to low voltages between scans. When all scans are analyzed, fragment data with a strong molecular ion peak are produced. In the same analysis information is provided for molecular weight determination and for fragmentation confirmation of structure.

14.6. AUTOMATED MASS CHROMATOGRAM QUANTITATION

Most LC/MS systems allow simultaneous display of UV signal and LC/MS TIC and SIC. They can be displayed with a time offset so that both types of chromatograms can be directly compared. Both can be integrated with peak detection integration using internal and external standard quantitation against calibration runs of known standards. The LC/MS data have the advantage of target compound analysis as in GC/MS quantitation. Known target compounds can be identified and quantitated using their primary and secondary fragment masses.

Impurities can be identified using the growing library databases for tentative identified compound searching. Once identified, the impurities can be added to the known compound list and quantitated to provide information on impurity levels. Information of this type would be unavailable from even photodiode array UV detectors.

14.7. FUTURE OF LC/MS

Liquid chromatography offers tremendous potential for analyzing nonvolatile polarized and ionized materials often found in real-world situations. These materials can be separated and analyzed with little or no sample purification, extraction, or derivatization. Ion spray offers the capability of analyzing these compounds without having to make severe alterations in LC conditions to accommodate the mass spectral detector. It also promises both molecular weight and structural interpretation through fragmentation data. Once MS detector price decreases further, the library databases increase to cover nonvolatile compounds, and the tuning capability of the detector becomes automated, these systems will appear on every laboratory research bench.

APPENDIX A

GC/MS TROUBLESHOOTING QUICK REFERENCE

This section is designed to assist in troubleshooting system problems. It is not meant to replace systematic troubleshooting and routine maintenance. A systematic approach to problem solving is always better. Keeping this in mind, in this appendix commonly seen problems, possible causes, and suggested treatments are listed. Appendix B contains a list of common background contaminants.

I. GC INJECTOR PROBLEMS

Problem 1. Peaks broaden and tail.
    Cause a: Poor column installation causing dead volume in the injector.
        Treatment: Reinstall column in injector. Check seal at ferrule. Check insertion depth. Ensure a good column cut.
    Cause b: Solvent flashing in hot injector.
        Treatment: Reduce injection speed on hot injectors and if possible reduce injector temperature. If you are using sandwich injection, reduce solvent plug to 0.5 μL.
    Cause c: Incorrect injector temperature control.
        Treatment: Typically set injector temperature 20°C lower than solvent boiling point and keep column at solvent boil-

ing point. Hold column at initial temperature until injector has finished heating.

Cause d: Septum purge line is plugged.

Treatment: Check that the septum purge flow is ~0.5 mL/min. Change the septum purge frit or adjust the needle valve.

Cause e: Injector not being purged properly after splitless injection.

Treatment: For splitless injection, the vent flow should be 70 mL/min, and the injector should be switched to the split mode 0.5–1.5 min after injection.

Problem 2. Tailing sample peaks for active components.

Cause a: Active sites in the injector insert or liner.

Treatment: Change or clean the injector insert. Silanize it, if necessary.

Cause b: Active sites or degraded phase in column.

Treatment: Remove the front 15 cm of the column and re-install. If retention times are changing or cutting the column does not help, replace the column.

Cause c: Injector not hot enough for higher boiling compounds.

Treatment: Increase the injector temperature and lower injection speed. Check that the graphite ferrule is free of cracks and the septum support is tight.

Problem 3. Low response and tailing of high-boiling-point compounds.

Cause a: Injector is not hot enough to vaporize high boilers.

Treatment: Increase injector temperature.

Cause b: High levels of column bleed are masking component peaks.

Treatment: Condition column or change to a high-temperature column if condition does not help. Consider changing to a bonded phase column if problem continues.

Cause c: High levels of silicone is coated on ion source surfaces.

Treatment: Clean the ion source.

Cause d: Interface/ion source not getting to adequate temperature.

Treatment: Change the manifold heater.

II. GC COLUMN PROBLEMS

Problem 4. Leading sample peaks.

Cause a: Column overload due to excess amount of component injected.

Treatment: Dilute the sample or do split injection.

Cause b: Degradation of stationary phase.

Treatment: Change the column. Change to a bonded phase column.

Cause c: Carrier gas velocity too low.

Treatment: Increase carrier gas flow rate.

Problem 5. Poor chromatographic resolution.

Cause a: Column temperature or program not optimized.

Treatment: Modify method by changing temperature ramp's segment slopes. (See GC methods development in Chapter 6.)

Cause b: Carrier gas flow rate not optimized.

Treatment: Decrease carrier gas linear velocity.

Cause c: Column not capable of the separation.

Treatment: Change to a more polar column. Change to a capillary column with a higher plate count.

Cause d: Stationary phase has degraded.

Treatment: Replace the column.

Problem 6. Peak size changes from run to run.

Cause a: Leaking or partially plugged syringe.

Treatment: Check visually that the syringe is pulling up sample. Remake Teflon seal around the autosampler syringe needle or flush the syringe with solvent. Heating the syringe in a hot injector may help if it is plugged; otherwise, replace the syringe.

Cause b: The septum leaks.

Treatment: Replace septum regularly. Ensure that the septum nut is tight.

Cause c: Improper column installation in injector or column inlet leak.

Treatment: Check installation of column in injector and tighten the capillary column nut.

Cause d: Sample is absorbed by active surfaces in injector or column.

Treatment: Change injector insert and deactivate it if necessary. Remove front 5 cm of column if it is a capillary column or replace the column.

Cause e: Sample is incompletely vaporized in the injector.

Treatment: Increase the injector temperature or the maximum programmed temperature of the injector.

Problem 7. Peak splitting, especially low boilers.

Cause a: Sample is flashing in the injector simulating two injections.

Treatment: Lower injector temperature. Use sandwich technique for splitless injection.

Cause b: Column temperature program starts before injector is heated.
Treatment: Increase initial column hold time until injector has reached its maximum temperature.

Cause c: Solvent plug in injector.
Treatment: Decrease solvent plug to 0.5 μL or eliminate if possible.

Problem 8. Extra peaks in chromatogram.

Cause a: Septum bleed, particularly during temperature programming.
Treatment: Use high-temperature, low-bleed septum. Ensure that septum purge flow is 0.5 mL/min.

Cause b: Impurities from sample vials such as plasticizers.
Treatment: Confirm by running solvent blank with new syringe. Change to certified sample vials and keep samples refrigerated.

Cause c: Impurities from carrier gas.
Treatment: Install or replace carrier gas filters.

Cause d: Contaminated injector or GC pneumatics.
Treatment: Remove column from injector and bake out at elevated temperature with a purge flow of at least 20 mL/min.

Cause e: Impurities in sample.
Treatment: Confirm by running a blank or standard run.

Problem 9. Retention times shift in chromatogram.

Cause a: Unstable carrier gas flow controller/regulator.
Treatment: Check pneumatics for leaks. Replace flow controller/regulator if necessary.

Cause b: Column contamination or degradation.
Treatment: Condition or replace column.

Cause c: Leaks at septum or column to injector connection.
Treatment: Replace septum regularly and check that the septum nut and the capillary column nut are tight.

III. MS VACUUM AND POWER PROBLEMS

Problem 10. High vacuum pump will not turn on. No pressure reading.

Cause a: Filament is burned out. No filament voltage.
Treatment: Shut down vacuum. Replace filament.

Cause b: System leak. Rough pump cannot reach starting vacuum.
Treatment: Find and repair system leak.

Cause c: Rough pump will not turn on.
Treatment: Check rough pump power. Replace pump oil. Replace pump.

Cause d: Turbo pump rotor has seized.
Treatment: Turn in turbo pump for replacement.

Cause e: High vacuum gauge cathode burns out.
Treatment: Replace cathode tube.

Problem 11. Cannot reach operating vacuum (10^{-6} torr).

Cause a: Contaminated fore or diffusion pump oil.
Treatment: Look for background increase in previous TIC. Replace fore pump oil. Have diffusion pump oil replaced by service representative.

Cause b: Analyzer contaminated by diffusion pump oil.
Treatment: Shut down mass spectrometer. Disassemble. Clean quadrupole rods with methylene chloride, acetone, or methanol. Buy a mass spectrometer with functional butterfly valves.

Cause c: Major air leak around column fitting into interface.
Treatment: Replace column ferrule and reseat compression fitting.

Cause d: O-ring around analyzer housing is not seating.
Treatment: Push down on analyzer housing cover with vacuum on. There should be a change in rough pump sound as vacuum increases. Replace housing gasket.

IV. MS SOURCE AND CALIBRATION PROBLEMS

Problem 12. With repeller at maximum, calibration gas 502 cannot be found.

Cause a: Source is dirty.
Treatment: Shut down vacuum. Clean ion source and lens.

Cause b: Mass axis is badly out of calibration.
Treatment: Autotune. If 502 peak is still off scale, calibrate 131 to 69, then 264 to 131. The 502 peak should be on scale. Calibrate 502. Recheck the 131 and 69 masses.

Cause c: Detector showing loss of sensitivity/burnout.
Treatment: Increase EM voltage to 3500. If still no 502 peak in the calibration gas, replace the detector.

Problem 13. Calibration gas 69 peak position moving. Poor 70 mass resolution.

Cause a: System grounding problem.
Treatment: Ground analyzer unit to control interface. Float the computer ground.

Cause b: Ion source is dirty.
Treatment: Shut down system. Clean the source.

Problem 14. No calibration gas peaks.

Cause a: Calibration gas valve is not open.
Treatment: Open calibration gas valve.

Cause b: Calibration gas solenoid valve stuck open. All calibration gas evaporated.
Treatment: Have solenoid replaced. Put fresh PFBTA in the calibration gas vial.

V. SENSITIVITY AND MS DETECTOR PROBLEMS

Problem 15. Analysis sensitivity has decreased.

Cause a: Background has increased.
Treatment: Check column bleed, septum bleed, pump oil, and ion source contamination.

Cause b: Detector needs replacement.
Treatment: EM voltage is over 3500 to see calibration gas 502. Replace detector.

APPENDIX B

SOURCES OF
GC/MS BACKGROUND
AND CONTAMINATION

The GC/MS is an extremely sensitive instrument. However, achievement of this kind of sensitivity is background dependent and requires elimination of all common sources of contamination.

Essentially, two kinds of background can interfere with trace-level GC/MS analyses:

1. *General background contamination,* such as column bleed, hydrocarbons, phthalate plasticizers, and so on, will generate a large TIC signal during the analytical scan and decrease the sensitivity level for detecting target compounds.

2. *Specific ions in the background* will interfere with a single-ion or extracted-ion chromatogram. For example, significant 164 background might be present when trying to detect low levels of 2,4-dichlorophenol. This type of problem is less common than general background contamination. Typically, a single ion or an extracted ion can be chosen that does not appear in this background.

The easiest way to determine if the background is permanent is to lower GC temperatures to 50°C and run a scan to see if the background decreases. If it does, the background is probably due to column bleed, septum bleed, contaminated pump oil, or leaks of various kinds.

In all instances where the background is determined to be coming from the analyzer and not eluting from the gas chromatogram, the system should be shut down and the source cleaned. If this does not eliminate the problem, shut down the system and dip the rods, washing with methanol or methylene chloride to remove contaminants. A permanent background is defined as background that is at approximately the same level regardless of GC temperatures.

The other source of problems, contaminated samples, tend to give more discrete chromatographic peaks and specific mass fragments. These samples can be cleaned using SFE or SPE cartridge or GPC columns before injection.

The GPC columns separate by size and release smaller molecules before the larger, polymeric material. They are very good for removing road-tar-like materials from extracted samples. Although getting the road tar off the column may be a problem, generally, if it can be dissolved, it can be eluted.

The SFE or SPE cartridge columns are true chromatography columns. They can be used to do class separations of materials. Using windowing techniques and standards, one can work out methods for purifying the materials of interest from either polar or nonpolar contaminants. This technique is described in *HPLC: A Practical User's Guide* (see Appendix D). Finally, their cost is low and if contaminated and contaminant cannot be washed out, they can be discarded.

Following is a list of some common contaminant mass ions:

Mass Ions	Compounds	Source of Origin
18, 28, 32, 44	H_2O, N_2, O_2, CO_2	Air leak
28, 44 CO, CO_2	Hydrocarbon fragments	—
31	Methanol	Lens-cleaning solvent
43, 58	Acetone	Cleaning solvent
69	Fore pump fluid	Saturated trap pellets
69, 131, 219, 254, 414, 502	FC43 (PFTBA)	Calibration gas leak
73, 207, 281, 327	Polysiloxanes	Column bleed
73, 207, 281, 149	Polysiloxanes	Septum bleed
73, 147, 207, 221, 295, 355, 429	Dimethylpolysiloxane	Septum breakdown
77	Benzene or xylene	Cleaning solvent
77, 94, 115, 141, 168, 170, 262, 354, 446	Diffusion pump oil	Improper shutdown of pump heater

Mass Ions	Compounds	Source of Origin
91, 92	Toluene or xylene	Cleaning solvent
105, 106	Xylene	Cleaning solvent
151, 153	Trichloroethane	Cleaning solvent
149	Plasticizer (phthalates)	Vacuum seals damage
14 amu spaced peaks	Hydrocarbons	Saturated trap pellets, fingerprints, pump fluid

APPENDIX C

GLOSSARY OF GC/MS TERMS

Base peak   The most intense ion fragment in a compound's spectrum under a given set of experimental conditions.

Capillary zone electrophoresis (CZE)   A separation technique based on movement of ionized compounds through a capillary tube filled with buffer toward a high voltage of the opposite polarity. Separation is based on the compound's size and charge potential.

Carrier gas   Gas used to sweep volatile materials from the injector through the GC column and into the detector.

Chemically induced (CI) ionization   Ionization in a MS source in which a diluting gas, such as carbon dioxide, is added to the analysis sample. The diluting gas, being in higher concentration, is ionized first and transfers this ionization to the sample at a low energy, forming a more stable molecular ion. Used in molecular weight determination.

Column   A packed tube filled with coated, absorptive stationary phase particles used to achieve GC separations.

Data/control system   The "brains" of the GC/MS system, which programs the system components, controls MS scanning and lens, and acquires and processes the data from the detector.

Detector   A device that produces a voltage change in response to a change in the composition of the material in its flow cell.

Differential pumping   An arrangement in which two chambers connected by a small orifice, such as an MS source and analyzer, have two pump connections through different-diameter exhaust tubes. Capable of providing different pumping rates and vacuums in the two chambers.

Direct insertion probe (DIP)   A metal probe with a slanting flat surface that allows sample to be inserted through a vacuum port directly into the ionizing electron beam in the MS ion source.

Efficiency factor   A chromatography resolution factor that measures the sharpness of peaks. Controlled by carrier gas nature and flow rate, particle size, coating thickness, and column diameter and length.

Electrode   A source of electrons for ionizing samples. *See also* Filament and Ring electrode.

Electron-induced (EI) ionization   Sample ionization in an MS source by bombardment with 70-eV electrons from a filament. This is a high-energy ionization leading to fragmentation of the original molecular ion.

Fast atom bombardment (FAB)   Ionization for a nonvolatile sample. Suspended in glycerol, it is placed on a DIP tip and inserted into a stream of heavy metal ions in the source. The matrix explodes, vaporizing the ionized sample, which is repelled into the analyzer.

Filament   Metal plates that connect to the ion source and release a stream of ionizing electrons when a voltage charge differential is applied.

Fourier Transform GC/MS (GC/FT-MS)   A separation technique in which a GC sample is ionized in an evacuated chamber, is held in place by a cyclonic trapping voltage, is excited to a higher orbit by a "chirping" multifrequency signal, and transmits an RF signal characteristic of all the masses present. Transformation of this multifrequency signal allows plotting of intensity vs. *m/z* spectra with very high sensitivity at each chromatographic point.

Gas chromatography (GC)   Separation technique in which the volatile analyte is swept by a carrier gas down a column with packing coated with an absorbing liquid. Differential partition between the two phases by sample components leads to band separation and elution into a detector.

Injector   A device used to move a sample in an undiluted form onto the head of a column.

Internal standard   A compound added during the last dilution before sample injection in equal concentration to all analyzed samples. Its purpose is to correct for variations in sample injection size. It also can be used to correct for variations in peak retention times.

Ion trap detector (ITD)   A desktop mass spectrometer that ionizes and holds the ionized sample in a circular electromagnet until swept with a DC/RF frequency signal that releases the ionized sample into the ion detector.

m/z   A symbol for mass divided by charge, measured in atomic mass units or daltons. The *x* axis for mass spectrum indicating that an MS spectra is dependent on both the mass and the charge on the fragment ion.

Matrix blank   A quality control, matrix only sample analyzed to show levels of target compounds present before spiking with standards.

Matrix spike   A QC sample required for 5% of all sample analyzed. A matrix blank is spiked with all standards at a level within the analysis range and checked for recovery of standards.

Molecular weight   Summation of the weights of all the elements in a molecule expressed in atomic mass units or daltons. In MS, the weight of the molecular ion in EI ionization.

MS/MS GC/MS system   A tandem, triple-quad system for study of MS fragmentation mechanisms. The second analyzer is a collision cell used to further fragment ions separated in the first analyzer for analysis in the third analyzer.

On-column injection   GC injection directly onto the head of the column; used to avoid loss of sample in the injector due to thermal breakdown. Has problem of nonvolatile contamination of column.

Pascal (Pa)   A measure of pressure equal to 7.5×10^{-3} torr (mm Hg). Commonly used in Europe, but also used by some U.S. manufacturers.

Quadrupole analyzer   Mass spectrometer analyzer base on four circular rods held in a hyperbolic configuration and swept with a variable-frequency DC/RF signal allowing selection of individual mass fragments.

Qualifiers   A major fragment peak in a compound's spectra; with the target mass, used to confirm the identity of the compound. Either the qualifier's mass or height relative to target mass can be used.

Reagent blank   A first blank run to indicate the cleanliness and the capability of the laboratory to run samples. Reagent water is spiked with all standards and subjected to full analysis conditions.

Resolution equation   A measure of a column's separating power. It combined retention, separation, and efficiency factors into a single equation that shows their interactions.

Retention factor   A column resolution factor measuring how separation is affected by the residence time on the column. Controlled by temperature and carrier gas pressure.

Retention time   The length of time a compound stays on the column under a given set of experimental conditions.

Ring electrode   The central electrode of an ion trap used to hold ion fragments in circular orbits until the time to elute them into the detector.

Roughing pump   The first pump in a vacuum system. It is used to reduce pressure initially from atmospheric pressure to a low pressure that can serve as a starting point for the high-pressure pump. Currently usually a mechanical rotary vane pump.

SCAN   An MS operation mode in which the amount of each mass unit is measured by continuously changing the DC/RF frequency on the quadrupole. Mass can be scanned low to high or high to low. The latter leads to less intermass tailing and more accurate relative-height measurements.

Separation factor   A column resolution factor controlled by the column's chemistry and temperature. Changes in this factor result in shifting of relative peak positions.

SIC   Single-ion chromatogram. Chromatogram produced by displaying the ion current produced versus time for a given mass (m/z). It can be produced by operating in a single-ion mode or extracted out of a scanned fragment database.

SIM   Single-ion monitoring. The mass spectrometer measures one or a few specific masses. Since fewer measurements are made than in SCAN, they are made more often with a proportional increase in sensitivity.

Spectra   A plot of signal intensity, measured in volts, versus ion fragment m/z, measured in atomic mass units, for a given MS scan or range of scans. The data are usually summed around unit masses and presented as a bar graph of intensities relative to the base peak.

Supercritical fluid chromatography (SFC)   A column separation technique using pressure/temperature control to convert a gas into a fluid that is used as the mobile phase for liquid/solid chromatography. Sample recovery is made by releasing the pressure to turn the mobile phase back into a gas.

Surrogate   A standard compound added in known amounts to all processed samples. Its purpose is to detect and correct for sample loss due to extractions and handling errors. Usually it is a deuterated or other labeled derivative of an analyzed compound not normally found in nature.

Target compound quantitation   Quantitation based on identifying a compound by locating its target and qualifier ion fragments. Once identified, the target ion signal strength is compared to known amounts of standards to determine the amount present.

Target ion   A compound's MS ion fragment chosen to identify and quantitate the amounts of the compound present in mixtures of standards and unknowns.

Temperature ramp   A gradual, controlled increase of temperature with time. It is used in combination with holds and other ramps in building an oven temperature program for resolving compounds on a GC column.

TIC   (1) *Total ion chromatogram.* A chromatogram produced by measuring the total ion current from the mass spectrometer versus time. A TIC data point represents a summation of all mass fragments present at a given time. (2) *Tentatively identified compound.* A compound, found in the chromatogram of an unknown, not listed as a target compound, internal standard, surrogate, or known compound. It is referred to library search and a reasonable number of matching compounds are reported.

Time-of-flight GC/MS (GC TOF/MS)   Chromatographic technique in which the MS detector analyzes effluent by ionizing it with a pulse of electrons and identifying mass fragments by the time they take to travel a flight tube and reach a detector. LC/TOFMS is becoming popular in the analysis of charged biochemicals, proteins, and DNA restriction fragments. Ionization occurs through pulsed laser energy.

Torr   A commonly used measure of pressure or vacuum equal to 1 mm Hg, or 133.32 Pa.

Triple-quad GC/MS/MS   A tandem quadrupole system in which a gas chromatogram feeds a mass detector with three quadrupole units in series. The second quadrupole as a holding and collision cell in which fragments separated in Q1 can interact with a heavy gas, xenon, and further fragment for separation in Q3. Used primarily for studying fragmentation mechanisms.

Turbomechanical pump   High vacuum pump that uses a series of vanes mounted on a shaft. They rapidly rotate between stator plate entraining air molecules, dragging them out of the evacuated volume. The "turbo pump" operates like a jet engine to evacuate the mass spectrometer to the high vacuum needed for operation (10^{-6} or 10^{-7} torr.).

APPENDIX D

GC/MS SELECTED READING LIST

JOURNALS

1. *Journal of the American Society of Mass Spectroscopy*
2. *Journal of Analytical Chemistry*
3. *LC/GC Magazine*
4. *American Laboratory*

BOOKS

1. F. W. McLafferty and F. Turecek, *Interpretation of Mass Spectra,* 4th. ed., University Science Books, Mill Valley, CA, 1993.
2. J. Throck Watson, *Introduction to Mass Spectrometry,* 2nd. ed., Raven, New York, 1985.
3. W. McFadden, *Techniques of Combined Gas Chromatography/Mass Spectrometry: Applications in Organic Analysis,* Wiley-Interscience, New York, 1973.
4. R. R. Freeman, *High Resolution Gas Chromatography,* 2nd. ed., Hewlett Packard, Palo Alto, CA 1981.
5. D. Ambrose, *Gas Chromatography,* Van Nostrand Reinhold, London, 1971.
6. M. C. McMaster, *HPLC: A Practical User's Guide,* VCH Publishers, New York, 1994.

INDEX